DISCARD

MATHEMATICS
IN FUN
AND IN EARNEST

Other Books by
Nathan A. Court

COLLEGE GEOMETRY
MODERN PURE SOLID GEOMETRY

MATHEMATICS
IN FUN
AND IN EARNEST

by NATHAN ALTSHILLER COURT

THE DIAL PRESS   NEW YORK

THIRD PRINTING, JUNE 1962
DESIGNED BY WILLIAM R. MEINHARDT
PRINTED IN THE UNITED STATES OF AMERICA
BY THE HADDON CRAFTSMEN, INC., SCRANTON, PA.

TO
David, Lois, and Ellen

Preface

Towards the end of the last century ("fin de siècle") the idea was afloat that the planet Mars was inhabited. Venturesome spirits were anxious to convince those Martians that their neighboring plant Earth, too, is peopled with intelligent creatures. Those Earthians, or Terrestrians had the urge to tell their hypothetical neighbors that they are not only interested in the inhabitants of the red planet but would also like to "hear" from them.

The surest way to bring about those desirable ends would be to construct, say, in the Sahara desert a gigantic geometrical figure of the kind used to prove the Pythagorean theorem. If such a figure powerfully illuminated were to be flashed into the sky at an appropriate time, it would surely attract the attention of the Martians and induce them to reciprocate in kind. For mathematical propositions are universal and eternal verities and therefore familiar to all intelligent creatures everywhere.

The Martians did not receive that signal: the project never got beyond the talking stage. But the story serves as "documentary evidence" of the attitude of the educated public towards mathematics as recently as half a century ago.

New conquests of science in general and of mathematics in particular have created a different intellectual climate which caused many preconceived notions to be abandoned, many a cherished myth to be given up.

The mathematician came to the conclusion that his science is a human enterprise, beset with all the foibles inherent in man's handiwork, but also resplendent with his creative power, his imaginative sweep, radiant with his love for beauty. The mathematician still feels that he is justified in his claim that Mathematics is the brightest jewel in the intellectual crown of mankind.

Most of the essays assembled in this book attempt to mirror this "new look" or new outlook of mathematics. They were written over a considerable period of time, mainly during the last decade or so. The author was privileged to be able to share his reflections with others, from the speaker's platform, over the air, and through the printed page of the periodical press. The more than favorable reception which was consistently accorded those utterances provided the incentive for collecting them within the covers of the same book.

Although these writings met with the approval of very competent judges, they were not composed for the benefit of the experts in the field. The author had primarily in mind the cultured lay reader whose intellectual curiosity impels him to try to keep abreast of the times, and on the other hand the professional whose field of special interest is more or less removed from the domain of mathematics. On that account care has been taken to avoid technical mathematics, and where recourse has been had to it, its scope does not surpass the High School level. And even that part may usually be skipped without necessarily losing the trend of the argument at hand.

By the nature of their origin, each of these essays is complete in itself. From the point of view of the reader this has the advantage that the book may be read section by section in any order that may be found interesting or convenient, not necessarily in the order adopted in the book. On the other hand this independence of the various sections from one another, and also their restricted size account for the fact that some topics are discussed in more than one place, although usually from a different angle. This is particularly true about the axiomatic method in mathematics. However, the fundamental importance of this subject, not only in mathematics, makes it bear repetition quite gracefully. The reader who would be interested in following up any given topic may be helped by the cross references in dicated in the text at the appropriate places and stated explicitly at the end of each chapter.

The lay reader would not be surprised to see that mathematicians are concerned with the history and philosophy of their subject, with the relation of mathematics to social problems, etc. But he is rarely led to suspect that the practitioners of the most "mysterious" of the sciences find within their subject room for recreation, for play, to use a simpler word.

It seemed to this writer that, to be complete, the picture of mathematics should also comprise something to the "lighter" side of this discipline. Moreover, the reader may perhaps be induced to try his own hand at playing some game of the mathematician. Should he yield to such a temptation he may be surprised to find how fascinating such a game may turn out to be. It may even happen that he, the reader, may be amused by this writer's clumsy, no doubt, and decidedly unorthodox attempts at fictionalizing, or dramatizing some geometrical propositions.

If some serious minded reader would come to the conclusion that the author does not always treat the earnest topics with quite the traditional dignity (synonym—stiffness) becoming such a subject, he may be perfectly right. But I do not propose to apologize for this misdeed. I would rather hope to win that reader over to my own credo:

Mathematics in earnest should be fun,
Mathematics in fun may be earnest.

The University of Oklahoma N. A. C.
Norman
June, 1958

Contents

Chapter I   MATHEMATICS AND PHILOSOPHY

Chapter II   SOME SOCIOLOGICAL ASPECTS OF
                 MATHEMATICS

Chapter III   THE LURE OF THE INFINITE

CONTENTS 15

CHAPTER VII   MATHEMATICS AS RECREATION

Acknowledgments

It is a great pleasure for me to express my indebtedness to my friends and colleagues for all I learned from then in our informal, private, and after dinner conversations, and for the opportunity I had to clarify my own ideas that find their expression in the pages that follow. I also thank those of them who read in manuscript some sections of this book or supplied me with references in the field of their competence.

Dr. Duane H. D. Roller earned my sincere gratitude for having read, carefully and attentively, a considerable part of the manuscript, and for his willing and obliging help as curator of the DeGolyer Collection of History of Science and Technology.

Sophie Court read intelligently and criticized mercilessly the entire manuscript, from beginning to end. She corrected all of the typescript and all the printed proofs with skill and devotion. To try to thank Sonia would be to belittle both her and her help.

Last but not least I wish to record my thanks to Clarkson N. Potter of The Dial Press, without whose initiative and sustained interest this book may perhaps not have seen the light of day.

N. A. C.

MATHEMATICS
IN FUN
AND IN EARNEST

Chapter I

MATHEMATICS AND PHILOSOPHY

1 · SOME PHILOSOPHICAL ASPECTS OF MATHEMATICS

Introduction. The historical relation of Philosophy and Mathematics is a matter that a mathematician may point out with some legitimate pride. Philosophy, as it is understood in our Western world, is the creation of the ancient Greeks. The Greek "love of wisdom" included the study of nature as well as the inquiry into the forms of human relations, that is to say it embraced all learning. Only later did different branches of philosophy break away from their parent and form independent disciplines. Physics, for instance, did not emancipate itself until sometime in the sixteenth century, and psychology only during the nineteenth century. Mathematics, however, was never a part of philosophy. It was recognized by the Greek philosophers as an independent intellectual pursuit from the very start. The history of mathematics in Greece runs parallel to the history of philosophy itself.

A · The Nature of Mathematics Mathematics has its roots deep in the soil of everyday life and is basic in our highest technological achievements. We use mathematics when we count the lumps of sugar for our breakfast cup of coffee, we use mathematics when we build our houses, erect our lofty skyscrapers, when we construct our wonderful printing presses and our imposing bridges, our mysterious radios, our supersonic airplanes. At the same time mathematics is reputed to be, and actually is, the most abstract, the most hypothetical of sciences. Let us illustrate this statement by an example. The Greeks considered what they called "perfect" numbers, that is, numbers which are equal to the sum of their divisors. The number six is such a number, for $6 = 1 + 2 + 3$. Again $28 = 1 + 2 + 4 + 7 + 14$ is a perfect number, and 496 is another example of such a number. Notice that all three examples offered are even numbers. Whether there are odd perfect numbers remains an open question. What is certain is that no such number has even been found. However, this circumstance has not deterred the mathematician from studying odd perfect numbers and proving theorems concerning them. In other words, the mathematician is ready to make the statement: "If there are odd perfect numbers, they will exhibit such and such characteristics."

B · The Unity of Form and Number The representation of numerical data in graphical form has achieved wide acceptance. The daily press, and periodicals in general, often have recourse to this device when discussing various phases of our economic life, like the fluctuations of commodity prices, population data, changes in the size of crops through the years, etc. Graphs are used by governmental agencies, by industrial and commercial concerns to render account of their activities. Curves representing data obtained in scientific experiments abound both in professional journals and in learned tomes.

This very efficacious union of number and picture, or form, is one of the applications of analytic geometry, or

Cartesian geometry as it is called after its inventor René Descartes (1596-1650).

However, the ambitions of analytic geometry in the use of number go much farther. This branch of mathematics replaces straight lines, circles, and other curves, by algebraic equations, and does the same for cylinders, spheres, ellipsoids, and other surfaces. By manipulating those equations analytic geometry obtains the geometrical properties of those curves and surfaces. Thus geometrical reasoning is replaced by algebraic operations, pretty much as the use of algebra takes the place of arithmetical arguments in the solution of problems which we in our earlier school days tried to puzzle out by arithmetic.

How far can this union of algebra and geometry be pursued?

The invention of projective geometry, early in the nineteenth century, brought forward the idea that geometric properties may be divided into two kinds, those of measure and those of position. The length of the circumference of a circle is a question of measure, but that this circumference cannot have more than two points in common with a straight line has nothing to do with measurements and is a property of position.[1] Now in the middle of the nineteenth century Karl Staudt (1789-1867) showed that the entire domain of projective geometry, which domain deals with questions of position exclusively, may be developed with complete independence from the notion of measure. On the other hand, the classical analytic geometry essentially presupposes a unit of length and therefore measurements. But this serious difficulty has not divorced number from geometry. Even before the work of Staudt had appeared, a suitable apparatus for the analytical treatment of projective properties was already in use.

But algebra has no exclusive rights to the exploration of geometry. The invention of the calculus was to a large degree inspired by geometrical problems. The calculus, in its turn, applied its great powers to further the study of geometry, and during the nineteenth century created the vast domain of differential geometry.

Furthermore, during the present century projective concepts were introduced into differential geometry, and thus the new doctrine of projective differential geometry came into being. The union between number and space instead of growing weaker has become so intimate that E. J. Wilczynski (1876-1932), one of the founders of Projective Differential geometry, could declare categorically: "Every problem of mathematical analysis (i.e., of the study of number) has a geometrical interpretation, and every problem of geometry may be formulated analytically."[2]

The philosophical implications of this union of number and space has not escaped the notice even of those who witnessed the birth of analytic geometry. The notion of number and the idea of space seem so far apart, qualitatively so different that the correspondence between algebra and geometry revealed by Descartes' invention is philosophically as far reaching as it is unexpected. This correspondence goes to show that the various concepts which we elaborate starting with different kinds of perception may not be as far apart as their origins would imply. Furthermore this correspondence may be indicative of the "unity of knowledge" or of the unity of the external world, although the aspects under which we both perceive and conceive that world may seem to us to be quite different.

C · The Dimensions of Space　The union between analysis and geometry sheds a brilliant light upon the question of dimensionability of space. A solid having three of its points fixed cannot move. If only two of its points are fixed, the solid is free to rotate about the line joining the two points, and we say that the solid has one degree of freedom. If only one of its points is fixed, it has two degrees of freedom. A solid none of whose points is fixed, has three degrees of freedom. Such is, in brief, the intuitive origin of our belief that space has three dimensions. On the other hand, in analytic geometry a point in the plane is associated with two numbers, x and y, the coordinates of the point, and vice versa, a pair of numbers is interpreted as a point in the plane. In space of three dimensions a point has three coordinates x, y, z, associated with it, and three numbers represent a point. We thus obtain another intuitive corroboration that our space has three dimensions.

However, analytic considerations lead us much further. Analytic geometry also shows that four numbers, a, b, c, d, determine a definite straight line in space, that is to say, the four given numbers may be interpreted as a definite straight line. If only three of these numbers are assigned, and the fourth is allowed to vary, the corresponding straight line will cover a surface; if only two of the four numbers are fixed, the corresponding line will form what is called a congruence, if just one of the numbers is fixed we obtain a complex of lines, and finally, if all the four numbers vary, the straight line fills space. A similar story may be told about the sphere, for a sphere is also determined in analytic geometry by four numbers. We are thus suddenly confronted with the fact that the dimensionality of space is a relative question, namely relative to the element with which we want to fill space. Our ordinary intuitive space is three-dimensional with respect to points, and four-dimensional with respect to lines, or spheres. Under the impact of these ideas space loses its majestic rigidity, its fixed and immutable form of a ready container, since it depends for its essential characteristic upon the element we want to fill it with. The mathematical importance of this conclusion is closely rivaled by its philosophical significance.

Another trend of thought opened by analytic geometry is the four-dimensional point-space. One number fixes the position of a point on a straight line, two numbers determine a point in the plane, and three numbers determine a point in our three dimensional space. By analogy four numbers may then be interpreted as a point in four-dimensional space. Of course, we have no intuition of such a space. But that has not prevented the mathematician from accumulating, by means of his analytical machinery, a vast number of theorems dealing with curves and surfaces in four-dimensional space. More than that, he sees no good reason why he should stop at four dimensions. Any n numbers may be interpreted as a point in an n-dimensional space, and the use of equations with n variable enables the mathematician to study n-dimensional space along the same lines along which he studies the three dimensional space which is so dear to our intuition and in which we profess to feel so perfectly safe. Later on we may say something about the "utility" of such studies. For the present we would only insist on the intellectual side of this creation. The world of perception, the world of intuition furnishes the suggestion of a three-dimensional geometry. But under this impulse the mathematician spins a new thing which is patterned after the old one, but which has nothing to do with experience,—just a product of pure intellect. The bearing that such a performance has upon the theory of knowledge and the sources of knowledge is obvious.

As a fitting conclusion to the discussion of the unity between number and space we may consider the question: What is it that makes such a unity possible? If it be granted that there is a complete equivalence between analysis and geometry, as Wilczynski so staunchly maintains, what is the profound common residue that accounts for this relation?

A discussion of this question may have important consequences. If we could find out precisely what characteristics of geometry enable us to identify it so closely with number, we may, in the process, discover the conditions which any science must satisfy in order that it may be identifiable with mathematical analysis. These sciences, if there be such, would immediately secure the powerful succor of this wonderful mathematical discipline, and their progress would proceed by leaps and bounds. On the theoretical side such an investigation would further enlighten us to the nature of knowledge itself. But even if these expectations should turn out to be too optimistic, the question limited exclusively to analysis and geometry is important enough.

D · Postulational Mathematics Mathematics is reputed to "prove" the propositions it advances. Now what does a mathematical proof accomplish? It shows that a mathematical proposition that has been proved is a logical consequence of one or more mathematical propositions which were already admitted to be true before. These latter propositions, in their turn, are logical consequences of other propositions admitted to be true, and so on. But this chain cannot be receding forever. We finally arrive at a proposition which we admit as being in no need of a proof, as being "self-evident", that is to say we arrive at an axiom or axioms of the mathematical science considered. Euclid started his geometry by laying down the self-evident propositions which he classified as axioms and postulates, and then derived, by the use of logic, all the propositions of his famous Elements. The development of mathematics during the nineteenth century taught us a great deal about the role of postulates and axioms in the body of a mathematical science.

In the first place a little reflection makes it clear that the "self-evidence" of the postulates—this term will be used to include both axioms and postulates—is a luxury we can readily dispense with. Indeed, any logical deduction made from the postulates will be logically valid, as long as we admit the postulates to be valid, regardless of whether this validity of the postulates is based on "self-evidence" or is just a convention. This is so obvious that it would seem useless to insist upon it. But it remained hidden from the mental eyes of mathematicians and philosophers alike, until the latter part of the nineteenth century.

The possibility of the conventional character of the postulates of mathematics is only one side of the medal. Let us now focus our attention upon a definite branch of mathematics, say, geometry. We define a polygon, for instance, as being made up of triangles, and a triangle as being made up of straight lines, and so on. But here again we must come to a point where we cannot reduce our terms to other terms more simple or more familiar unless we agree to turn in a circle, as the ordinary dictionaries actually do. We may, for instance, define a straight line as the shortest distance between two points. But this simply assumes that the notion of distance is more familiar to us than the notion of a straight line. We are thus confronted with the same situation as to the basic terms of geometry as we were before when we tried to trace the validity of our propositions back to its origin. We must admit some terms of geometry to be "self-explanatory", to be in no need of any further elucidation beyond an appeal to our common knowledge, to our intuition. A further careful study of the postulates and propositions of, say, plane geometry, reveals the astonishing fact that these propositions do not involve any of the physical properties of points and lines beyond the relations of these elements to one another which are stated explicitly in the postulates of the science, as for instance that two points determine a line. The terms "point", "line", thus become mere words to designate things which satisfy the postulates of geometry, but devoid of any other meaning.[3]

Now if we combine the arbitrariness of the fundamental postulates of geometry with the lack of meaning of the terms involved in these postulates, we obtain a strange picture of geometry, or of any other mathematical science, for that matter. It is a perfectly coherent logical structure about things the meaning of which we ignore, beyond a certain number of formal relations, and of propositions which inform us about nothing more than that one statement is true, if another statement be granted to be true. "A is true, if B is true." What an enchanted world for the mathematician! What an enormous freedom for intellectual endeavor, without any restraints or impediments! If a mathematician takes a notion to create a mathematical science, all he has to do is to set up a group of postulates to suit his own taste, postulates which he by his own fiat decrees to be true, and involving things nobody, including the mathematician himself, knows much about, and he is ready to apply formal logic and spin his tale as far and as fast as he will. If any human being ever was entitled to lose his head in a fit of megalomania, it was the mathematician at the opening of the present century. And he did lose his head. All mathematics—just nothing, but a child of the mathematician's brain, a structure in which not only the plan and design, but even the foundation, even the very material it is built of—nothing but a product of his imagination. The mathematician felt himself to be the great master of creation, and so secure in his greatness that he could afford to poke fun at himself and at his own expense, as in the famous phrase: "A mathematician never knows what he is talking about or whether what he says is true."[4]

The mathematician was fortified in his conceit by the signal success of his new conceptions on the structure of mathematics. It has given him an understanding of the different branches of his science that he could not have reached otherwise. It has shown him a new way of erecting new disciplines with a firmness and security that he had not experienced before. Furthermore he found out that he has a model to offer for others to imitate. C. J. Keyser (1862-1947) has shown, in the *Yale Law Journal* for February 1929, in an article entitled: "On the Study of Legal Science", how the postulational method may apply in this domain, apparently so remote from the mathematical world. R. D. Carmichael (1879) suggests that the physicists ought to abandon attempts to base physics on "self-evident", that is to say, anthropomorphic postulates and terms, if they are to succeed in their efforts to make physics a mathematical science. The Frenchman J. Rueff advocates the extension of the postulational method to the social sciences in his book, *Des Sciences physiques aux sciences morales* (Paris, 1922).

The arbitrariness, the lack of specific meaning of the terms employed in a mathematical science, has a great advantage. The whole theory of geometry, for instance, developed from point and line as basic elements, may be applied to any other two terms, if these two terms fit the fundamental postulates. The whole doctrine thus becomes a mold, a form ready to receive contents as soon as a set of terms may be found to fit the mold. The meaning of the propositions will be different with the different sets of terms used, but the structure of the doctrine will remain the same. C. J. Keyser calls such a doctrine a "doctrinal function". The projective geometry in three dimensions elaborated for the terms point, line, plane is perfectly valid, if we interchange the terms point and plane. The individual propositions are different, but the mold as a whole remains intact. A Russian mathematician A. A. Glagolev in his doctoral dissertation devised a scheme which makes the geometry of all the circles in the plane fit the mold of the point geometry of the ordinary three dimensional space. The potential applicability of these ready molds, of these doctrinal functions cannot be foretold, but their very existence is a philosophical achievement of no mean magnitude.

We may now return for a moment to the question of the unity of number and space, that we discussed before. If there is a complete equivalence between analysis and geometry, that can readily be explained by assuming that these two doctrines are two different forms of the same doctrinal function. To be sure, the fundamental postulates in analysis and in geometry are very different. But this is no serious objection. The same mathematical doctrine may be developed from several different sets of postulates. If we start with the postulates of one set, the postulates of the other set become propositions in the body of the doctrine developed, and vice versa. It is therefore possible that a set of postulates and a set of undefined terms may be found, and a doctrinal function developed from them, which would become in turn analysis and geometry, if we replace the undefined terms of this doctrinal function, in turn, by the specific terms of analysis and of geometry.

E · The Question of Consistency Postulational mathematics is one of the great conquests of the human spirit. However, the paradise of freedom and sovereignty which the mathematician arrogated to himself in connection with this wonderful creation has early enough proved to be a precarious place to dwell in. The postulates of a mathematical science may be laid down arbitrarily. The rest of the doctrine is developed by pure logic and the test of its validity is that it must be free from contradictions. Such a result cannot possibly be attained, if the postulates themselves involve a contradiction. Hence the first and cardinal requirement a set of postulates must satisfy is that it be consistent with itself, that it be free from inner contradictions. It is therefore the first duty of the mathematician to verify that his postulates possess this indispensable property. But how is that to be accomplished? There is the *a posteriori* way: the mathematician may begin to derive propositions from his postulates, and if he encounters no contradiction, he may assume that his postulates are consistent. The question then arises, how many such propositions does he have to derive to be entitled to the desired conclusion? ten? a hundred? a thousand? a million? It is clear that such an *a posteriori* proof cannot be satisfying. The necessary thing is a test applied directly to the postulates themselves, an *a priori* proof. The discouraging thing about the situation is that the shrewdest minds among mathematicians have not been able to devise a logical criterion by which to test the consistency of a set of postulates. The tests actually used are not of a logical, but of a physical nature. If the postulates are true propositions about a concrete set of objects, the postulates are judged to be consistent, on the assumption that no existing thing can have two properties which are contradictory. Little stress is laid on this point by writers on the subject. This method of procedure is very significant. It expressed the fundamental belief that logical consistency is identical with natural consistency. It makes the consistency of nature to be one of the foundations, one of the cornerstones of the mathematical edifice. This is already a far cry from the dictatorial powers of the mathematician, as we have considered them before.[5]

F · The Empirical Origin of the Axioms   The restriction that the postulates of a mathematical science must be consistent, and that this consistency can be tested only in an empirical way is a severe blow at the principle of arbitrariness of these postulates. Nor is this all. The most famous book that exhibits the postulational method of developing a mathematical science is David Hilbert's *Foundations of Geometry*. In spite of the theoretical freedom of choice of the postulates, Hilbert happened to choose the same postulates as did Euclid over two thousand years before him. Hilbert's work is more systematic in this respect, it reflects all the acquisitions made in this domain of thought by the intervening centuries, but the postulates are essentially the same, and therefore the resulting geometry is Euclidean geometry. The number of different sets of postulates which have been worked out as a basis of geometry is considerable. But all of them are equivalent and arrive at the same geometry. Why do not these mathematicians use the freedom that is theirs, that they have won at the cost of an immense effort? The answer is very simple and at the same time of basic importance in the understanding of the role of mathematics in relation to epistemology and to the other sciences. The postulates of Euclid gave us a geometry which works in the world we live in (leaving out the questions raised by the theory of relativity), a geometry which is practical, which tells us something about this world, a geometry that fits the other branches of human knowledge. Furthermore, the fact that the geometry deduced from these postulates is applicable to the physical world shows that those postulates themselves have a physical basis, that they are empirical laws, refined and abstracted laws, but laws derived from experience, just as the very notions of point and line are abstracted from the physical world. Let us take the time to give one illustration. Given a triangle, it is not possible to draw a line which would cut its three sides between the vertices. The older mathematicians made use of this property, although this is not a consequence of the postulates of Euclid. They took this property to be so obvious that it required no proof, and the reader, no doubt, would be in agreement with the mathematicians on this point. Now this "self-evidence" is nothing else but empirical. Modern mathematicians when they are trying to be complete in the statement of their postulates, include this property in their list.[6]

The manner in which the postulates of Euclid have been derived from experience is a psychological problem which has been discussed by mathematicians like Poincaré and Enriques, to mention only two. The classification of the postulates as empirical laws makes of the whole body of geometry a physical science, the most perfect physical science, if you will, but a physical science nevertheless. David Hilbert in an address delivered before a congress of naturalists said: "Indeed, geometry is just that part of physics which describes the relations of position of solids to one another in the world of real things."[7] And this by the famous author of the modern classic on the foundations of geometry!

Geometry is a more perfect science than the other physical sciences because the number of postulates, or of fundamental laws, if you prefer, is small, and on that account we succeeded in collecting all, or nearly all those which were necessary for the erection of the stately geometrical edifice.

G · The Worth of Deductive Reasoning   Within the limits of logical consistency the choice of the postulates of geometry is arbitrary and the mathematician is free to make any choice that would suit his fancy. But he sacrificed his freedom, first subconsciously, and then deliberately and knowingly. Or rather he has freely chosen the Euclidean postulates, upon the suggestion of the environment he lives in, and the rest has been done by logical deduction.

The question may be raised, in passing, about the value of this deduction. Strictly speaking, deductive reasoning cannot teach us anything that is not contained implicitly in the premises. Hence all deductive reasoning is a tautology, a roundabout way of saying that A is A. Does this apply to geometry? Are all propositions of geometry quite obvious to anyone who masters the first postulates of the science?

When Newton first came across a copy of Euclid's *Elements* he read in it a page or two here and there and ended up by throwing the book under his bed with the contemptuous remark: "Now, all this is too obvious." In the course of a lecture on mathematical analysis the writer's professor remarked about Henri Poincaré: "This man handles analysis with such dexterity that he really believes the subject is easy." And the lecturer went on to reinforce his remark by telling his listeners that Poincaré, when a student at the famous École Polytechnique of Paris, attended the course of lectures in analysis without a book and without a notebook. Apparently analysis may be too easy, too. Now these may be just legends. But whether the stories are true or not, Newtons and Poincarés are mighty few and far between, and even for those privileged darlings of fortune the going gets rough ultimately, after a certain level is reached.

We have to admit that while the deductive reasoning is not capable of producing anything that is not in the premises, such efforts make explicit what those premises contain and imply; and the results thus elicited, as far as the student, the knower, is concerned, are quite new and form a valuable addition to his knowledge. The nature of the game is such that the postulates become overwhelmed by the mass of consequences to which they themselves give rise.

The postulates of geometry make the existence of a triangle logically possible, but they do not compel its consideration. A triangle is a further product of human experience or of human imagination and, as such, is an addition, in a way, to what is contained in the postulates. The consideration of the circle circumscribed about the triangle is a further step in this creative direction, and the comparison of two triangles another invention of the human mind. From triangles we pass to polygons, from the consideration of one circle to a group of two circles, of three circles, of the infinite number of circles passing through two given points. This creative capacity of the human mind that accounts for the fruitfulness of the passage from a triangle to a polygon, from one circle to a group of circles is nothing else but induction. The familiar process which mathematicians call "generalizing" is induction, quite comparable to the induction which is the foundation of the physical sciences.[8]

H · Imagination and Imitation Geometry, like mathematics in general, is a combination of basic postulates, which are physical laws, and the creative imagination of the human mind, the two elements being joined together by deductive logic. What is the source, what is the impulse of the creative capacity of the human mind, in connection with mathematics? The answer is twofold: observation and imitation. The world around us includes many objects having the approximate form of triangles, circles, and so on. On a higher plane, many advanced theories of mathematics are due to questions raised

by the physical sciences. The invention of the calculus, for
instance, is largely due to such questions. We have spoken
before of the basic terms, the so-called undefined terms of a
mathematical science, and we pointed out that they need not
have any meaning beyond satisfying the basic postulates of
the science. But when this is actually the case, if there is no
physical picture attached to these terms, and the imaginative
capacity of the mind is left to its own resources exclusively,
progress in the science is slow and laborious. Take four-
dimensional geometry. We admit that there is no direct intuition attached to it, and its propositions are but theorems concerning equations in four variables. Nevertheless, we insist on
using the language of geometry, because this language is
suggestive, it points towards avenues of investigation and thus
helps the imagination which would be very much hampered
without such aid.

This four-dimensional geometry may also serve as an
example of what we called "imitation." We create mathematical theories which are based on intuition. But once we
have such a theory, we may create another similar to it, but
for which we have no empirical model to follow. Thus we
have four-dimensional geometry, non-Euclidean geometry,
etc. On a smaller scale this may be observed in the everyday
work of the mathematician. Suppose the mathematician comes
across the problem of finding the path of a point which moves
so that the ratio of its distances from two fixed points is constant, and he finds that this path is a circle. He will immediately ask himself, what that path would be, if instead of the
ratio, the product of the distances were constant, or the sum,
or the difference. He is thus led to the study of curves which
may have little in common with the circle.

It may be interesting to note that these two forms of
activity of the human mind, namely the following of the world
outside and the imagining of new things which have no counterpart in the external world, far from being distinct and

separate, keep on crossing each other's path, intertwining to the extent of making it impossible to tell them apart. The history of mathematics is full of cases where problems taken from the physical world have given rise to mathematical theories, and conversely, the creation of the imagination of the mathematician later found its application in physical problems. The conic sections were little more than a mathematical pastime with the Greeks and are common structural forms in our time. The imaginary numbers, as their very name indicates, were created by the mathematician almost in spite of himself, with little faith even in their legitimacy. For the last half century or so these fancy numbers have become an indispensable part of the mathematical theory of electricity, and a valuable tool in electrical engineering.

The great French mathematician, J. L. Lagrange (1736-1813), proposed the problem of determining the surface of least area which would pass through a given curve. In 1873 Joseph Plateau, (1801-1883), the blind physicist of the University of Ghent, described an experimental way of realizing such a surface by means of soap bubbles made of glycerin water. Such bubbles tend to become as thick as possible at every point of their surface, and the surface thus becomes as small as possible, that is a minimal surface. But a mathematical solution of the problem was not forthcoming. As late as the first quarter of the present century competent mathematicians were willing to venture the opinion that mathematics may not have developed far enough to cope with this problem. In 1931 a complete solution of the problem was published by a young American mathematician, Jesse Douglas (b. 1897).

I · Conclusion It will be fitting to conclude this discussion with the following quotation from the address of David Hilbert referred to before. Says Hilbert: "For the mathematician there is no ignorabimus, neither is there one for the natural sciences, in my opinion. The philosopher, Auguste

Comte (1798-1857), said one day—in order to point out a problem that is certainly insoluble—that science will never succeed in piercing the secret of the chemical composition of the celestial bodies. A few years later this problem was solved by the spectral analysis of Kirchhoff and Bunsen, and one may now say that the far away stars are important physical and chemical laboratories of a kind that have not their like on earth. In my opinion, if Comte has not succeeded in pointing out an insoluble problem it is because there is no such. Instead of falling into a senseless agnosticism we ought to adopt the following slogan: 'We must know, we will know.' "

2 · GEOMETRY AND EXPERIENCE

A · Origins of Geometrical Knowledge Students who gather for their first lesson in geometry already know a good deal about the subject. They are familiar with certain shapes that textbooks on geometry call parallelepipeds, spheres, circles, cylinders, which the students would call boxes, balls, wheels, pipes. Notions such as point, line, distance, direction, and right angle are quite familiar and clear to them, in spite of all the difficulties learned mathematicians profess to encounter when they try to clarify or define these concepts.

The question arises: how was this store of knowledge gathered, how was this information acquired? The empiricists maintain that geometrical knowledge is the result of the experience of the individual in the world surrounding him.[9] However, the universal acceptance of the basic properties of space lead the apriorists to the conclusion that these spatial relations are innate, that they constitute a fundamental characteristic or limitation of the mind which cannot function without it or outside of it. The invention of non-Euclidean geometry by Lobačevskiĭ[10] has done considerable damage to the solidity of the apriorist armor but has not eliminated the debate between the two schools of thought.

During the present century the eminent French sociologist Emile Durkheim (1858-1917) advanced an intermediate thesis. The source of our geometric knowledge is experience. However, at a very early stage of civilization this individual experience is pooled and codified by the group, owing to social necessity and in order to serve social purposes. Our basic geometric knowledge is thus a social institution. It is this social function of geometry that accounts for the fact of its universal acceptance, for the inability of the individual to act contrary to it, for the mind to reject it.

It is universally agreed that the actual experience of living is the basic factor in the process of accumulating information of the kind that we call spatial or geometrical. This in turn amounts to saying that we come into possession of this information through our senses. Such being the case, the question naturally comes to mind, which of our senses is it that performs this function?

The sense of hearing helps us acquire the notion of direction. To a lesser degree this is also true of the sense of smell. The sense of taste need hardly be mentioned in this connection. The sense of sight and the sense of touch remain. It does not take much effort to see that these two senses play the dominant part in the shaping of our geometrical knowledge.

B · The Sense of Touch The sense of touch, considered in its broader aspect of including also our muscular sense, informs us of the shape of things. It is also our first source of information about distance. By touch we learn to distinguish between round things and things that have edges, things that are flat and things that are not flat. It is the sense of touch that conveys to us the first notions of size. This object we can grasp with our hand, and this other cannot be so grasped; it is too big; this object we can surround with our arms, this other we cannot; it is too big.

These examples imply measuring, and the measuring stick is the size of our hand, the length of our arm, and, more generally, the size of our body. The whole environment that we have created for ourselves in our daily life is made to measure for the size of our body. That the clothes we wear are adapted to the size of our body and our limbs goes without saying. But so is the chair we sit on, the bed we sleep in, the rooms and the houses we live in, the steps we climb, the size of the pencil we use, and so on, without end. We take it so much for granted that things should fit our size that we are startled when they fail to conform to the adult standard, as, for example, in the children's room of a public library where the chairs are tiny and the tables so very low. The legendary robber Procrustes, of ancient Greece, had his own ideas about matching the sleeper and the size of the bed. He made his victims occupy an iron bed. If the occupant was too short, he was subjected to stretching until he reached the proper length. If, on the contrary, the helpless victim was too tall he was trimmed down to the right size, at one end or the other. Hebrew writers placed this famous bed in Sodom, and it was one of the iniquities that caused Sodom's destruction, by a "bombardment from the air."

In many cases the fact that things are made on the "human scale" may be less immediate but is no less real. The clock on the wall has two hands, whereas, strictly speaking, the hour hand alone should be sufficient. Owing to the limitations of our eyesight, we cannot evaluate with sufficient accuracy fractional parts of an hour by the use of the hour hand alone, unless the face of the clock was made many times larger than is customary. But then the clock would become an unwieldy object, out of proportion to the other objects around us made to the "human scale."

The comparison of the size of objects surrounding us with the size of our body is not just a kind of automatic reflex but is a deliberate operation as well. When, in the course of our cultural development, the need arose for greater precision in describing sizes and for agreement upon some units of length, we turned to our body to provide the models. The length of the arms and of the fingers, the width of the hand, the length of the body and of the legs all served that purpose at one time or another, at one place or another. The yard is, according to tradition, the length of the arm of King Henry I. The origin of the "foot" measure requires no explanation, and we still "step off" lengths.

C · The Sense of Vision   The sense of vision is the other great source of geometrical information. To a considerable extent this information overlaps the data furnished by the sense of touch. Sight informs us of the difference in sizes of objects around us. Sight supplements and extends the notion of distance that we gain through touch. Sight tells us of the shape of things, and on a much larger scale than touch does. But sight asserts its supremacy as a source of geometrical knowledge when it comes to the notion of direction. Moreover, sight tells us "at a glance" which object is closer, which is farther, which is in front and which is behind, which is above and which is below. Sight is supreme in telling us when objects are in the same direction from us, when they are in a straight line. When we want to align trees along our streets, we have recourse to sight. The fact that light travels in a straight line is one of the main reasons for the dominant position the straight line occupies in our geometrical constructs. Some learned persons will smile indulgently at the statement that a ray of light is rectilinear. The writer will, nevertheless, stick to his assertion as far as our terrestrial affairs are concerned, whatever may be true of light on the vaster scale of the inter-stellar or intergalaxian universe.

D · Metrical Geometry and Projective Geometry   Up to this point the geometrical knowledge that has been mentioned is the kind familiar to "the man in the street." Let us now turn to the systematic study of the subject, to the science of

geometry. Are both empirical sources of geometrical knowledge reflected in systematic geometry? Is it possible to classify
geometrical theorems on that basis?

If we examine Euclid, we see that he leaned heavily
towards tactile geometry, or the geometry of size. His main
preoccupation was to establish the equality of segments and
angles, to prove the congruence of triangles. The method of
proving triangles to be congruent consists in picking up one
triangle and placing it on the top of the other, which implies
that the moving triangle does not change while it is in motion.
This possibility of rigid motion was much insisted upon by
Henri Poincaré (1854-1912) and is now considered by mathematicians to be the characteristic property of the geometry
of size, or, to use the professional term, of metric geometry.
Euclid's is thus metrical geometry exclusively, or nearly so.
This is not at all surprising, since metrical geometry is the
geometry of action, the geometry that builds our dwellings and
makes our household utensils. The very origin of Euclid's
geometry is supposed to be connected with the parcelling out
of plots of land in Egypt after the recession of the flood waters
of the Nile.

Euclid did not know that his was metrical geometry. To
him it was just geometry, for he knew of no other kind. Neither
did his successors, in spite of the fact that they added to
Euclid's *Elements* a considerable number of geometric propositions which in their nature are visual and not metric. There
are numerous such propositions, some of them of fundamental
importance, in the collection of Pappus, a Greek author of
the third century of our present era. A systematic study of
visual geometry had to wait for a millennium and a half before
it found its apostle and high priest in the person of the French
army officer Jean Victor Poncelet (1788-1867), the father
of projective geometry.[11]

Consider any geometrical figure, say a plane figure
(triangle) F (Fig. 1), for the sake of simplicity, and let S

be a point (representing the eye) not in the plane of figure F. Imagine the lines joining every point of figure F to the point S. Now, if we place a screen between S and figure F, everyone of these lines will mark a point on the screen and thus we obtain a new figure F′ in the new plane, the image of figure F.

If we compare the two figures F and F′ we notice some very interesting things. The figure F′ in general will be different from F. It has suffered many distortions. If A, B, are two points in F, and A′, B′ are their images in F′, the distance

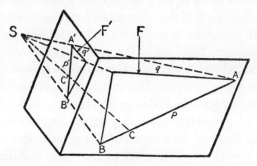

Figure 1

A′B′ is not equal to the distance AB, as a rule, and may be either smaller or greater than AB, and this alone deprives the figure F′ of any value in the study of the figure F from a metrical point of view. There are, moreover, many other distortions of various kinds. But some characteristics of F always reappear in F′. Of these the most important is that a straight line p of F has for its image in F′ a straight line p′, and consequently any three points A, B, C of F that lie on a straight line in F will have for their images in F′ three points A′, B′, C′ that also lie in a straight line. If two lines p and q are taken in F, their images in F′ are two straight lines p′ and q′, but the angle p′q′ is not equal to the angle pq, as a rule, and may be either smaller or larger than pq. In particular, the images of two parallel lines are not necessarily parallel, and the images of two perpendicular lines are not necessarily perpendicular.

If we call figure F′ the projection of figure F from the point S, we may say that projection preserves incidence and collinearity. The systematic study of projective geometry, or visual geometry, is the study of those properties of figures that remain unaltered by projection, just as it may be said of metrical geometry that it is the study of those properties of figures that remain unaltered in rigid motion.

From the point of view of the theory of knowledge it is of great significance that the distinction between tactile geometry and visual geometry was not noticed by either philosophers or psychologists. Only after the patient labors of mathematicians created the doctrine of projective geometry did the distinction come to light. The credit for having pointed out this distinction goes to Federigo Enriques (1871-1946), late Professor of Projective Geometry at the University of Rome.

In the study of the sources of our geometrical knowledge too little attention is accorded to our own mobility, to our ability to change places. Even the range of our knowledge due to touch is considerably increased by our ability to move our arms. In connection with our visual information our mobility is of paramount importance. To mention only one point, the shape of an object depends upon the point of view, from which it is observed. It is our ability to change places that makes it possible for us to eliminate the fortuitous features from our observations.

E · Conflicting Testimony of the Senses   As has been mentioned before, our tactile and visual information do not cover the same ground, but they overlap to a considerable extent and thus complement each other. But do they always agree? If a person drives his car over a stretch of straight road, he observes that the road is of the same width all along. He knows it to be so by comparison with the size of his car and by comparison of the size of his car with his own size; in other words, it is a tactile fact. Now, if he turns around and looks at the road just traversed, he sees "with his own eyes" that the road is getting narrower as it extends back into the distance and seems to vanish into a point. These two items of information on the same subject contradict each other. Which of them is true and which is false? Which of them do we accept and which do we reject? Above all, how do we go about telling which to accept and which to reject?

When one puts a perfectly good spoon into a glass of water, he sees that the spoon is unmistakably broken, or at least bent at a considerable angle. He takes the spoon out, and it is as good as it was before he put it in. He runs his finger along the spoon while it is in the glass and feels that it is straight as ever. But when he looks at it, there is no doubt that the spoon is bent; contradictory testimony of two different senses. Again the question arises, which of the two pieces of information do we accept, and on what ground do we make our choice?

A long time ago I read of a lake where the water was so clear that on a bright moonlit night it was possible to see the fish asleep on the bottom of the lake. Devotees of fishing would take advantage of this situation and go out in a boat, as quietly as possible, to the middle of the lake and then try to catch the fish by striking them with a harpoon. But simply to aim the harpoon at the spot where a fish was seen would spell disastrous failure. Successful practitioners of the sport would know the spot at which to aim, although the fish was seen to be elsewhere.

The moral of this fish story is of great importance. In the case of the road and in the case of the spoon we all repudiate the testimony of our eyes and accept the verdict of the sense of touch. We do so whenever the tactile and the visual testimonies are in disagreement. But why?

The answer to this puzzling question may be found in

the activity of man. Moreover, his activities are purposeful and must be co-ordinated so as to achieve success. Now, man's organs of activity, his hands, are also the main organs of touch. Man has thus developed a close coordination between his touch and his actions. At short range, he has implicit faith that his actions will be fruitful if he relies on the data furnished by touch. Visual data concern objects at a distance and serve well as a first approximation. They are good in most cases but are always subject to control and check. If light sees fit to indulge in such vagaries as reflection, refraction, and mirages, so much the worse for light. The fish story told above points to just that moral. Sight leads us to the fish. But if we want to act on it successfully, we must subject this information to the necessary correction as learned by touch. Otherwise we shall have no fish to fry.

3 · THE MIGHT AND PLIGHT OF REASONING

Introduction   Every inhabitant of this vast land of ours is aware of the fact that "in the city of Boston, the city of beans and of cod, the Lowells speak only to the Cabots, and the Cabots speak only to God." Mathematicians are just as exclusive a clique; they, too, speak only to their own kind. But the intellectual heirs of Pythagoras have gone those proud and haughty Bostonian families one better: they promoted God into membership in their own clan. A prominent British mathematician and astronomer figured out mathematically that God is a pure mathematician. And long before that, Plato decided that God (in His spare time), "geometrizes."

However, mathematicians are not very happy about the esoteric character of their science. In fact, they deplore it, for it causes them a great deal of embarrassment. When he has the opportunity of addressing a non-professional audience, the mathematician must leave behind, however regretfully, the field with which he is most familiar and move 'way out, on the fringes of his science, in order to find a terrain on which he can meet and commune with people who do not belong to his own fraternity. The adventure is alluring, but is also fraught with danger. If he does not go out far enough, he will bore his audience, and if he goes out too far he may be caught trespassing on somebody else's private territory and bore the experts, to say nothing about the danger of his "sticking his neck out." It is no simple task both to "satisfy the wolf and to save the sheep," to borrow a metaphor from the Russian peasants.

A · Reasoning and Psychology   Reasoning is by no means the exclusive prerogative of civilized man. Primitive man reasons too, and so do animals, for that matter. The question is only of degree. It is a standard anthropological method to try to understand man in his present state by studying him in his earlier stages of development. But this procedure is not suitable, if we want to analyze man's thinking processes. The features of human reasoning can best be discerned on samples where this reasoning has been checked and rechecked by successive generations of thinkers, as is the case with mathematical propositions.[12] A textbook on plane geometry includes samples of reasoning which are as good as any the human race is capable of producing.

Mathematicians not only use reasoning, but they also like to reflect upon this subtle art. They have contributed a great deal to the discussion of the logical aspect of reasoning, beginning, say, with Leibniz, to limit ourselves to modern times. During the present century mathematicians appropriated logic altogether and turned it into a branch of mathematics under the name of "Symbolic Logic."

Besides the logical aspect of reasoning there is also the psychological aspect. Cassius Jackson Keyser (1862-1947) said:[13] "Select a well wrought demonstration and examine

it. What can you say of it? You can say this: A normal human
mind is such that if you begin with such-and-such principles
or premises and with such-and-such ideas and if you combine
them in such-and-such order, it will find that it passed from
darkness to light—from doubt to conviction. Obviously such
a proposition is not mathematical; it is psychological—it
states a fact respecting the normal human mind." The same
idea was expressed even more pointedly by Henri Lebesgue
(1875-1941) who said: "Les raisons de se déclarer satisfait
par un raisonnement sont de nature psychologique, en mathématiques comme ailleurs." (The reasons for declaring one-
self satisfied with a reasoning are of a psychological nature,
in mathematics as in anything else). It seems the psychologists
have not come to grip with this problem. But they have approached the question of thinking from another side.

B · The Role of the Body in the Reasoning Process   Everyone knows that if you want to think you have to use your
head. Does any other part of the body participate in this
process? When a little boy, before I reached my 'teens, I
made an astounding discovery. One day, after school, before
I got ready to do my homework, I engaged in a hard running
game with some boys of the neighborhood. When I finally
yielded to the call of duty I was amazed that I could not find
an opening sentence for my composition, and that reading my
geography lesson instead was just as fruitless. I finally decided to work my arithmetic problems. But I could not solve
the problems. I was sure the world was coming to an end.
It did not, however, and I drew the conclusion that there must
be some connection between my mental efficacy and the physical state of my body. I know from repeated personal experience, just as everyone else does, that one may be too tired
physically to be able to read a book, or even a newspaper.

Ask the average man to explain to you the meaning of the
adjective "solid". He will tell you that it means something

strong, something substantial, and while doing so he will more likely than not clench his fist as tightly as he can.

Modern psychology has studied this subject in a methodical way. If you try to imagine a flying bird or a moving automobile, there is a tension in the muscles of your eyes and a tendency for you to roll your eyes in the direction of the imagined motion. When you imagine that you are bending your right arm, or that you are lifting a weight, your muscles become tense in the same way as though you were actually trying to do those things. If you imagine that you are counting one, two, three, etc., the tension in your speech musculature is the same as it is when you actually count aloud. On the other hand, one who is deprived, say, of the left arm is not able to imagine that he is lifting that arm.

It is important to emphasize that these are not mere assertions or "hunches" on the part of the psychologist. He has measured those tensions with instruments and has graphical records of his findings like those, say, which the weatherman has of the variations of the temperature during the day. His experimental evidence is just as solid as the evidence on which the experimental physicist bases his findings. The experimental psychologist is thus led to the inescapable conclusion that mental acts are performed not in the brain, or at least not in the brain alone, but that you perform those mental acts with your muscles, in a rudimentary way, to be sure, but your muscles come into play nevertheless. Their role is somewhat like the role played in theoretical mechanics by "virtual velocities" and "virtual work." Thus, in a paradoxical way, it may be said that we think with our muscles. Some psychologists go even further. They maintain that every part of our body participates in the process of thinking.

It is tempting to mention here an interesting analogy from the field of esthetics.

Music is a form of noise, "the most expensive of all disagreeable noises", as a very distinguished music hater once put it. As sound, music is directed at your ears. You surely will not be surprised if you are told that you usually listen to music not only with your ears but with your whole body. Few people can listen to music, particularly if the music is more or less familiar, without moving their body, or, more specifically, some part of their body, say, the head, or an arm, or a foot. When listening over the radio to a symphonic concert by a first rate orchestra, one is tempted to direct that body of performers, although knowing full well that at the other end of the line there is a competent conductor on the job. Strange as this behavior may be, one has a very good reason for engaging in the competition. One cannot derive all the enjoyment out of music unless one participates, so to speak, in its performance. This participation finds its expression in the more or less pronounced motions of one's body. By doing so you "feel yourself into the music". Students of the psychology of esthetics describe this attitude of the listener by the word "empathy." They insist that analogous things may be said about other forms of art, but we shall not go into that. Suffice it to point out here that according to this theory it may be said, paradoxically again, that "you appreciate beauty with your muscles."

C · *A Definition of Reasoning*   Having observed that thinking is helped or at least accompanied by muscular activity, we may raise the question: what is reasoning? What is it that we do when we reason? You need not be told that this is no mean question to answer. Perhaps some light may be derived from watching the reasoner at work.

Two men have an eight gallon keg of wine. They want to divide the wine equally between them. They have at their disposal two empty kegs of five and three gallons. How could they accomplish the division? One who is not addicted to reasoning may try it by the direct experimental method, i.e., by pouring the liquid from one container into the other. With sufficient patience and some good luck he may succeed.

One who is prone to reason will attack this well known riddle in a different way. He will still do the pouring, but only mentally. He may fill, say, the three gallon keg and make a note of the fact that he has five gallons left in the big keg. This is registering the result of a physical operation performed only mentally, and this implies that the reasoner is in possession of the information, acquired previously, that if you take three gallons from eight gallons there remain five gallons. Proceeding in this manner the reasoner will succeed in his task presumably much faster than the actual experimenter, especially if he is able to keep a record of the various pourings, either in his memory, or by using some mnemonical device, like, say, writing.[14]

The above description of the method of solving the riddle suggests that the reasoning involved consists of a series of physical operations performed in the imagination only, and that the performance of the operations mentally is made possible by the reasoner's knowledge, from previous experience, of the outcome of each individual operation.

Let us try the same scheme on another example. A block of wood in the form of a cube 3 x 3 x 3 inches is painted, say, blue. If the block were sawed up into one inch cubes, how many faces of each small cube would be blue?

Here again the question may be answered experimentally by actually sawing up the big block and counting the number of painted faces of the individual small blocks. But the reasoner may arrive at the outcome of the sawing without having recourse to the actual operation, relying for his answer upon his knowledge, that is to say, his previous experience with the cube. The reasoner will say that a small cube occupying a corner of the original block was a part of three of the faces of the big cube and will thus have three of its faces painted, and there are eight such little cubes. A small cube that was a part of an edge of the block but not placed at a vertex was part of two of the faces of the block and will therefore have two painted faces. There will be twelve such cubes. The six little cubes that were at the center of the faces of the block will have one painted face, and the little cube that occupied the center of the block will have no paint on it.

This example confirms our observation that to reason is to perform experimental work mentally, the outcome of each step in the chain of experiments being known to the reasoner from previous experience.

This definition of reasoning was given by the late professor of philosophy of the University of Pavia and editor of the renowned periodical *Scientia,* Eugenio Rignano (1870-1930), in his book: *La psicologia del ragionamento.* This work has been translated into French, German, and English, but it does not seem to have received the attention it deserves.

Let us consider the problem: If A walks to the city and rides back, he will require five and one quarter hours; but if he walks both ways he will require 7 hours. How many hours will he require to ride both ways?[15]

The problem may be solved in various ways. It will illustrate our point best to argue the case as follows: If A should make the trip to the city twice, walking one way and riding back, he would require for that $5\frac{1}{4}$ x $2 = 10\frac{1}{2}$ hours. Now two such trips are equivalent to one round trip on foot and a round trip riding both ways. But the former, we know, requires 7 hours, hence the latter trip will take $10\frac{1}{2}$—$7 =$ $3\frac{1}{2}$ hours. This reasoning confirms Rignano's definition so well that comments are unnecessary.

Rignano, himself, considers the following example: A pendulum clock is keeping good time in a given room. How will the clock be affected if it be transported into a room where the temperature is markedly lower than in the first room? The question may be answered experimentally by observing the clock in the new location. But one may reason out the answer, if he is in possession of some experimental facts, namely, the effect of temperature upon the length of a metal bar, and the relation of the length of a pendulum to the length of its swinging period. The reasoner so equipped will be able to say that the pendulum will become shorter in the cold room, and the shorter pendulum will swing faster, hence the clock will be fast.

However, as I mentioned before, the best place to check the validity of this definition of reasoning is in mathematics, and a textbook on plane geometry would do as well as anything.

Consider the proof, by superposition, that a triangle is congruent to a second triangle if two sides and the angle included between them of one triangle are respectively equal to the corresponding elements of the other. The steps in the proof are nothing else but physical operations performed mentally, "in the imagination." We put one vertex upon the other, and we know, by previous experience, that without changing the position of that point we still can revolve one triangle about that point. We revolve that triangle so as to make one side of it fall on the corresponding side of the other triangle, and so on. All these operations we perform mentally, and we are able to perform these successive steps because we know the outcome of each step from previous experience.

D · Applications of the Definition   The examples to which our analysis was applied were deliberately chosen for their simplicity, in order to bring out as clearly as possible the salient features of what reasoning is. However, the same features will be found if we examine reasoning on any level, no matter how abstract. The subject matter of the reasoning will in such cases be not the physical facts, but abstractions, symbols representing such facts and groups of facts. The methods of operation remain the same.

Let us now apply our definition of reasoning to some outstanding intellectual problems.

One of the questions that has preoccupied philosophers

through the ages is that of the rationality of the world we live in. How is it that the results of reasoning, the products of our inner intellectual effort are applicable to the external world and find their verification in it? Is it that reason pervades that world? If this be accepted as the answer, it is still possible that only a part of nature is rational and thus accessible to our mind, while the rest of the universe is irrational and thus completely closed to our intellectual perspicacity. The world would thus be divided into two distinct and mutually exclusive parts: The knowable and the unknowable.

This whole question with its mysterious profundity vanishes, if our description of reasoning is correct. If thinking about the external world is to perform a series of physical operations mentally, the outcome of each individual operation being already known from actual experience, then reasoning is simply returning to nature what we have learned about it directly. It is therefore not surprising that the results of correct thinking are in conformity with what happens around us. If the predictions of the astronomer concerning, say, an eclipse of the moon, come true, this is simply because each step in the reasoning which predicts the phenomenon is a verified fact of the physical world. We know the external world first and reason about it afterwards, and not the other way about.

Is it possible to learn to reason? Or better, is it possible to improve one's reasoning abilities? This is a pedagogical problem of basic importance. We all know that the native endowment regarding people's ability to reason varies, and the gamut of variation is enormous. It is the experience of any teacher that some of his pupils take to reasoning as naturally as a duck takes to water, while it is almost the exact opposite with others. Some teachers of voice maintain that anybody can learn to sing. Those singing enthusiasts will agree, no doubt, that it would take many years of training and a lot of sustained effort for some people to attain the skill in singing that others possess by accident of birth. But nobody doubts that the singing proficiency of a person can be improved by training. Is the same true about reasoning abilities?

Some decades ago a theory was current in pedagogical circles that training is not transferable from one domain to another. The learning skill acquired in one branch of knowledge does not help in mastering another. In particular, the practice of reasoning in one field carries with it no advantage when it comes to reason in another field. The proponents of this doctrine offered a lot of experimental data in support of their contention. In spite of that they could not win over the skeptics. In more recent years the original doctrine was modified. The reasoning ability acquired in one field is useful in another, but this transfer is not automatic. Students must be trained in acquiring skill in transfer, and "teaching for transfer" became a sound educational practice.

E · Pitfalls and Merits of Reasoning The original non-transfer theory is obviously in contradiction with what we have been saying about the nature of reasoning. If reasoning consists in mentally performing physical acts in some purposeful succession, the ability to do so should clearly be subject to improvement when practiced on any subject matter, and this improvement should have noticeable effects when reasoning is applied to a different field. But this transfer is indeed not automatic. The reasoner, in order to be successful, needs to know the result of each of his mental operations beforehand, from past experience. In other words, effective reasoning requires on the part of the reasoner a familiarity with the new situation, the knowledge of the facts in the case. It is therefore impossible to transfer reasoning ability acquired in one domain to another, if the subject matter of the new domain is not known to the reasoner. The point may be illustrated by the following example, which makes up in effectiveness what it lacks in depth. A charming young lady boasted before her chum that she had already had six marriage proposals. "And I am only eighteen," she added demurely. "That ain't so many," piped up her little brother Johnny, age eight, "that's just one proposal in three years."

On the authority acquired through many years of teaching the subject the writer can assure you that the little boy's arithmetic is absolutely impeccable. It is perhaps less certain that Johnny is quite familiar with all the facts regarding courting, proposals, marriages, etc., involved in the situation.[16]

Lack of familiarity with the domain to which the reasoning is applied is not the only trap into which the reasoner may fall. The chain of mental experiments may be so long that he may not be able to keep track of all the links and thus arrive at an erroneous conclusion. The reasoner must at all times be aware of this danger and check his results whenever there is an opportunity. The fellow whom we watched as he sawed that wooden block in his mind would do well, before he quits his job, to ascertain that the various little blocks that he produced mentally when taken together will account for the twenty-seven small cubes he expects to have.

He would also do well to check it in another way. He may count the number of painted faces on all the small cubes and see whether they add up to the $3 \times 3 = 9$, $9 \times 6 = 54$ little squares on the faces of the original block.

Another danger that lurks in the path of the reasoner is that he may overlook the fact that a new link that he brought in has a bearing upon the results he already admitted and that those results thus require some modification or adjustment. Here is an illustration. A man whose back was much stronger than his head dug a hole in the ground. When filling the hole up again he was very much perturbed that the hole could not hold all the dirt he took out of it. After much head scratching he finally discovered the reason for his predicament. "I must not have dug that hole deep enough."

Mistakes in reasoning would not be so disturbing if we could console ourselves that only inexperienced or poorly endowed individuals are the victims of those pitfalls. But unfortunately this is not the case. Sometimes we blame the social sciences when their arguments and predictions are off the mark. But we like to think that the so-called exact sciences are free of errors, and above all that mathematics is always correct. You may be surprised to learn that the mathematical literature includes a great many mistakes. Some of them are rectified rather promptly, others remain a long time unnoticed, and presumably a good many may remain undetected. Will all the mistakes in the mathematical literature ever be eliminated? This is a question that cannot be answered by "yes" or "no". What is more aggravating is that among the papers in which errors have been found are some that came from the pens of the bearers of the most illustrious names that adorn the annals of mathematics.[17]

In spite of these weaknesses, and many others, we are not likely to give up our mental experimentation for direct action, for various reasons.

It is much easier and more convenient to perform operations mentally. The mental process requires no equipment, no apparatus, no installations of any kind. Furthermore, it saves a great deal of time, to say nothing about expense.

What is even more important, the "thought experiment" has a much greater degree of generality than any physical experiment can have. If we superpose two material triangles we are tied by and to the two triangles at hand. When performing the same operation mentally, we can also mentally vary the two triangles and notice that the result of the operation is quite independent of the two triangles used. Thus we perform not one experienmt at a time, but a great many experiments in the same time, and are able to embody the results of all of them in one statement, in one proof.

The mental, instead of the material performance of the

experiments has also the merit of showing clearly the inter-dependence of the component parts of the "experiment" or "reasoning", a thing that would escape notice in the material execution of the experiment. After the cube we considered before is sawed up it is easy to establish that there are small cubes having paint on two faces, and others that have paint on three faces. But why are there twelve of the first kind and only eight of the second? The wielder of the saw has no answer for that, but the "reasoner" had no trouble account-ing for it. In fact, he could not have done his mental sawing had he not known this beforehand.

F · More Checks on the Definition   The mathematicians of the nineteenth century made a great contribution to human understanding when they invented postulational, or axiomatic mathematics. New branches of geometry that came into being during the early part of that century, non-Euclidean geometry among them, led mathematicians to the surprising conclusion that a mathematical science can be built by choosing arbi-trarily a set of objects or entities and by promulgating, just as arbitrarily, a set of rules or axioms which those entities are to obey. Those "undefined terms"[18] and "unproved prop-ositions" are the clay out of which the proposed science can be fashioned. Plane geometry has been built according to this model. The undefined terms are "point" and "line", and the "unproved propositions" are, roughly speaking, the axioms and postulates of Euclid.[19]

In principle, the choice of the undefined terms and of the unproved propositions for the building of the science of plane geometry is practically unrestricted. These terms and propositions need not be anything else or anything more than the creation of the human mind, a lucky product of the excited imagination, without any relation to or connection with the external world.

"The breath of life" that is supposed to pulsate in this

postulational clay is, of course, reasoning. Postulational mathematics begins by providing the material to which reasoning is to be applied. It thus conforms to the requirement stated before that we must have subject matter first and reason upon it afterwards. It may further be said that the postulates furnished provide enough "results of experiments known to the reasoner from previous experience" and thus satisfy this other requirement that would make it possible for reasoning to function in the way we said reasoning does. Thus, postulational mathematics seems to confirm, or at least does not conflict with our theory of reasoning.

The enthusiasts of postulational mathematics were outdone by a school of thought that has become known as logicalism. The exponents of this school blame the postulationists for granting too much. The logicalists maintain that in order to erect the whole edifice of mathematics nothing more need be assumed than the power to reason correctly, according to well defined rules. Weighty volumes have been written by foremost thinkers in support of logicalism.

As far as our present discussion is concerned logicalism flatly contradicts what we have attempted to present as the nature of reasoning. Contrary to what we have said, logicalism begins with reasoning and undertakes to produce the subject matter of mathematics as a result of it. Even if one would attempt to find consolation in the fact that the rules of logic which the logicalists begin by setting up may be taken to be a sort of preliminary subject matter, the discrepancy still remains wide enough to "give us pause," as Hamlet said.

But if we seem to disagree with logicalism, we may find comfort in the fact that we are in this respect fellow travelers of another, more recent school of mathematical thought known as intuitionism. The institutionists take sharp issue with logicalism. They categorically deny that mathematics is a corollary to logical reasoning. In fact, they reverse the entire situation and maintain that logical thinking is a by-product of mathematics, a technique developed in and for the study of mathematics, or any science, for that matter. Thus the attitude of the intuitionists towards reasoning comes so close to coinciding with the idea we have been discussing that one hardly could hope for a more striking confirmation.

You listen to an argument and then you declare: "I do not understand." What do you mean by that statement? In the light of what was said about the nature of reasoning you would mean that you do not visualize the chain of operations involved in the argument developed. And when you grasp the connection, you declare: "I see," with a sigh of relief, or a feeling of triumph, as the case may be.

If this interpretation is correct, it may help us to understand what Keyser and Lebesgue meant when they made the startling and disconcerting statement that the reasons for which we accept a logical argument are of a psychological nature. That "nature," according to our way of looking at reasoning, is the need to see the chain of mental experiments involved and to be sure of the outcome of each of them individually. If we can follow these steps, the argument is acceptable, and not otherwise.

G · Reasoning, Memory, and Invention   In what preceded I have tried to present in a plausible way what reasoning is, a definition which may have been repeated too many times already. We shall now mention briefly some other facets of the subject.

The reasoner has to have recourse to facts of experience which he learned previously. Obviously, he cannot make use of the necessary facts, unless he remembers them. Abel Rey lays great stress on the important role memory plays in reasoning and in the acquisition of knowledge generally. It is our memory which enables us to "fuse the past with the present in order to foresee the future."[20]

Think of any proof in plane geometry, say, the proof of

the Pythagorean theorem. Before you begin your argument you say that you draw such a line, or you join two such points, etc. How do you know which line to draw, or which points to join, or to do one rather than the other? This choice is invention. Even if you would argue that many false moves were made first, and that the lines drawn in the textbooks of plane geometry in connection with this theorem are the result of trial and error, all possibilities could not have been tried, for they are too numerous, and the element of invention still remains.

Let us consider another example. A cylindrical tube sealed at both ends stands on one end, on a table. The upper part of the tube is opaque, while the lower part is transparent and is filled with a liquid. How far into the opaque part of the tube does the liquid extend? All it takes to answer the question is to turn the tube upside down. We know, from experience, that the column of liquid will reach the same level regardless of the end the tube may stand on. If the liquid reaches above the middle line of the tube in one case, it will be just as much above that line in the other. "If equals are taken from equals, the results are equal." Does not that sound familiar? All this is good reasoning and conforms perfectly to the pattern we have been considering. But what made you think of turning the tube upside down? This is invention. It is an indispensable part of reasoning, but it is not reasoning, or at least it is a different aspect of reasoning. In the case considered another method, another invention could have served the purpose just as well. A plumber, under the circumstances, would have taken a different course. He would leave the tube in its original position and would drill a row of holes in it, beginning from the top. He would stop when the liquid would begin to flow out of the tube and would conclude that the level of the liquid was between the last two holes drilled. The plumber is inventive, too. But whichever method you might choose, you have to invent it. Abel Rey maintains that there exist positive, if obscure ties between intuition, invention, and the subconscious. Any invention, from the humblest and up to one which upsets the whole economy of human thought, is just an analogy which escaped notice up to that moment.

A French writer of the seventeenth century said: "Le choix des idées est invention." (The choice of ideas is invention.) How one goes about the business of choosing one's ideas, about the business of invention, is a question that preoccupied and baffled many of the greatest mathematicians. A most significant contribution along this line is due to Jacques Hadamard of the Collége of France.[21]

4 · PLANE GEOMETRY AND PLAIN LOGIC

A · The Impact of Non-Euclidean Geometry and of Projective Geometry The foundations on which Euclid reared his marvelous *Elements* endured for more than 2000 years. The first effective thunderbolt that struck these foundations originated in a remote Russian town, Kasan, located on the lower reaches of the river Volga, and was hurled at them in 1826 by an obscure professor of mathematics, Nikolai Ivanovich Lobačevskiĭ (1793-1856). Lobačevskiĭ's object was to prove that the parallel postulate of Euclid was not an obvious truth. In that he was eminently successful.[22] As a by-product, he wrought a change in our conception of the world we live in that has been compared, and rightly so, to the epoch-making achievement of that other Slav, Mikolai Kopernik, better known as Copernicus (1473-1543).

By a queer whim of history, about the time when Lobačevskiĭ meditated on his new geometry in Kasan, in another Russian town, Saratov, further down the Volga from Kasan, a young French officer of Napoleon's Grand Army of 1812, J. V. Poncelet, was whiling away his long solitary prison days in another kind of geometrical speculation. His lonely efforts were destined to become the foundation of a new branch of geometry and to form the contents of his famous

Traité des propriétés projectives des figures, which was published in 1822.

One of the strange ideas contributed by projective geometry is the principle that any of its propositions, in plane geometry, remains valid if we replace in it the points by straight lines and the straight lines by points. As a consequence of this "principle of duality," each theorem that is proved provides another theorem as well, which does not require a new, direct proof. The number of theorems is thus automatically doubled.[23]

The principle of duality was a great surprise. It was important enough to give rise to an acrimonious dispute over its paternity between two geometrical luminaries, J. D. Gergonne (1771-1859) and "the father of projective geometry," Poncelet himself.[24] Nowadays novices to the mysteries of projective geometry are confronted with this principle right at the start, as with a proposition which is practically self-evident. Those of the neophytes who shamefacedly confess that they do not grasp this idea quite clearly are assured by their elders that further progress in their studies will bring more light, and faith will be sure to follow. It does, usually.

The importance of the "windfall" that the principle of duality contributed to geometry is quite obvious. But the philosophic by-product which that principle entailed is even more far-reaching. We have a theorem dealing with certain entities—namely, points and lines. If this theorem remains valid when these entities are replaced by some others (in this case by lines and points, respectively), then our original theorem is not specifically a statement about points and lines. If we press this trend of ideas further, we are in the end confronted with the devastating question: What is it that we are talking about when we make our statements in geometry?[25]

B · The Formalist Approach to Geometry Non-Euclidean geometry and the principle of duality called into question the foundations of geometry and of mathematics in general. This was a much-discussed topic during the 19th century, both by philosophers and by philosophically minded mathematicians. Toward the turn of the century, the interest in these matters was greatly stimulated, among professionals and laymen alike, by the philosophical writings and lectures of Henri Poincaré (1854-1912) because of his towering scientific eminence, and perhaps even more because of his literary talent. These essays are excellent reading even today. The best English edition of most of Poincaré's contributions along these lines was prepared by G. B. Halstead.[26]

One outcome of the 19th century discussions was a deeper insight into and a more systematic use of the axiomatic, or formalist approach to mathematics in general, and to geometry in particular. In the case of plane geometry, the method consists in starting out with two kinds of objects named "points" and "lines" about which we profess to know exactly nothing. These objects have for us absolutely no other connotations than those bestowed on them by the propositions we explicitly formulate about them, and by which we are to be governed. These propositions are selected arbitrarily and declared to be true. When sufficient "axioms" have been accumulated, we are set up in business and are ready to start on the erection of the superstructure, with the help of the powerful lever of pure logic. The most highly regarded work along this line was done by David Hilbert (1862-1943). Thus, in this conception, plane geometry is just one grandiose creation of the human mind, one in which the senses and the sensory world have no part whatever.[27]

Imposing, even inspiring, as the edifice of the formalist may be, the obscurities of its starting point seemed to some to smack of sheer mysticism, and its proud aloofness from the world around us appeared to others to border on the absurd. But the actual heel of Achilles of this purely intellectual doctrine is that it suffers from an inherent intellectual weakness. The arbitrary choice of the fundamental axioms is subject to an obvious limitation: the axioms have to be logically consistent with each other. Hilbert labored for many years trying to produce a proof that the axioms of his *Grundlagen der Geometrie* satisfied this requirement. But all his persistent zeal and his enormous intellectual resources proved unequal to the task, though he could find some personal consolation in the proposition, proved by K. Goedel in 1931, that the "Grundlagen" *could not* yield a proof of its own consistency.[28] Georges Bouligand formulates the argument as follows:[29] "To find within a body of doctrine G a proof that G is consistent is impossible, for to accept the validity of such a proof is to concede to a part of G a special privilege: an abusive procedure, if the coherence of G as a whole is in doubt." Simple and obvious, David Hilbert to the contrary notwithstanding.

The shortcomings of formalism have brought out the limitations of the axiomatic method but have not impaired its value. Originated more than 2000 years ago in geometry, this method continues to lure other sciences by its undeniable advantages. Among the more recent conquests of, or converts to, the axiomatic method are the biological sciences.[30]

C · Role of the Knower The geometrical advances that were realized in the first third of the 19th century called into question the validity of the postulates of geometry as well as the nature of the entities it deals with. It was inevitable that sooner or later the instrument that geometry uses to manipulate these materials—namely, logic—should in turn be subject to scrutiny. What are the inviolate laws of logic? How and where have they acquired their infallibility? On what is based their tyrannical power over the mind of man?

And while we are in the questioning mood, would it not be appropriate to cast an inquiring eye on the manipulator of this powerful tool—the geometer himself? Does not the mental and physical make-up of the investigator have a bearing on the results obtained in the investigation? May not the Knower's knowledge depend on the Knower himself? Or, to put it broadly, is not the conception we make for ourselves of the world we live in influenced by the kinds of creatures we are ourselves?

Let us deal with the latter problem first. The questions of the dependence of our knowledge on our own physical and mental constitution are of rather recent origin. In the mental domain they were first considered by Kant. An adequate discussion of the entire problem would require knowledge of our nervous system that at present is not available. But once we raise these questions, the nature of the answer is beyond doubt.

When we look at an object, or at a landscape, and are not quite certain what we see, we turn our heads, or we move closer, or we walk around the object. Our knowledge thus depends on our ability to move—that is, on our physical structure. How utterly different this world of ours would be to us if we were immobile!

We explore our surroundings with our five senses (or is it six?). But what is so fixed and immutable about this number? Could we not have a larger number of them? The question is not quite as preposterous as it might seem at first. We have a sense for light. Why could we not have a sense for electricity? As matters stand now, the only way we can feel that mysterious stuff directly is to be shocked by it, sometimes to death, sometimes into health. If we want to make electricity accessible to our senses, in a less violent form, we resort to the expedient of transforming it into light. But we could conceivably have nerve ends that would convey to us the sensation of such electromagnetic waves directly. We know that our nerve ends that convey the sensations of high temperatures are different from those that convey sensations of low temperatures. That such an extension of our perceptivity within the domain of electromagnetic waves is not a physiological impossibility is attested by the fact that the visual spectrum of some animals reaches beyond the limits of the visual spectrum of man. For instance, insects, as a class, respond to electromagnetic radiations from both the ultraviolet and the infrared.[31]

A substantial extension of the range of electromagnetic waves directly perceptible to us through our senses would obviously materially affect our conception of the world around us. Radio astronomy would not have had to lag several thousand years behind optical astronomy.[32] At any rate, it certainly would have saved us all the time and all the effort that we had to spend, and still do, to study this form of energy by our indirect methods.

If we may find it difficult to talk about additional senses, it is easy to imagine that we might have been deprived of some of those we have. We all know unfortunates who are handicapped that way. Certainly for writers of science fiction such a conjecture is no trick at all—witness the story of H. G. Wells, *The Country of the Blind*. Those who write just science, pure and simple, should not have much trouble either. They know full well that we are blind, at least relatively, compared with other creatures.

The well-known limitations of our auditory perceptions offer occasion for analogous remarks.

D · Axiomatic Method   Federigo Enriques in his *Problemi della Scienza* of 1906[33] points out that the foundations of knowledge are more clearly discernible in knowledge that has already evolved than when it is still at its crude beginnings. This idea was taken up by Ferdinand Gonseth, author of *Les fondements de mathématiques, Les mathématiques et la realité, Qu'est-ce que la logique,* and so forth. The *Cumulative Index of Books in English* does not list any books by Gonseth. In order to find an answer to the question bearing on the nature of logic, Gonseth first subjected to a psychological examination the axiomatic method as applied to plane geometry.

A city is described by its plan in a schematic way. This schema usually furnishes information about the location of the streets, the public buildings, the transportation lines, and so forth, but has nothing to say about private residences or the location of the taxicab stations. The plan is thus only a simplified or summary description of the city.

The plan, or schema, is obviously incomplete, and additions may be made to it if necessary. The plan of the city may always be enriched, say, by marks indicating the location of the service stations of any enterprising oil company. Moreover, the things that are indicated on the plan are represented by conventional marks or symbols.

Thus the schema is a summary, symbolic and unfinished. The city that the schema represents may be said to be the exterior meaning or the *exterior significance* of the schema.

However, we may consider this schema by itself and for itself, without reference to the thing that it is supposed to represent. As such, the schema has its own reality and may be an object of study for its own sake. We may, for instance, examine the network of lines indicating the one-way streets or the pattern formed by the points marking the locations of the post offices and relate it to the similar pattern formed by the telegraph offices. We may even solve some geometrical problems that those figures might suggest. Of course, by these intrinsic considerations regarding the schema, we are diverting our attention from the plan's original purpose. On the other hand, such studies may well be undertaken in order to serve that very purpose with greater efficiency. The profound analogy between this example and geometry is so transparent and so striking that it can hardly be overlooked.

In order to accept the edge of a ruler as a realization of the abstract concept of a straight line, we must, in the first place, reconcile ourselves to the approximate character of this realization. But this is not enough. We must also be willing to forget that the correspondence between the concept and its realization holds only macroscopically and that it vanishes completely when the edge of the ruler is put under a macroscope. In other words, the realization of the concept is only summary. What we said of the straight line can obviously be repeated about any other concept used in geometry. In the light of our example with the plan of a city, we say that *rational geometry is a schema of ideas whose exterior significance is to be sought in a certain natural structure of the physical world*. We are thus quite far from the much quoted quip of Bertrand Russell: "In mathematics we don't know what we are talking about, or whether what we say is true."[34]

Pursuing our analogy between rational geometry and the plan of a city, we may say that to set up our geometrical schema means to conceive, in a summary and schematic fashion, a set of simplified notions and a number of relationships among them. To reason intrinsically on this schema means to render explicit the consequences implied in those relationships. In other words, to develop the reality of schema is to set up a system of statements having the value of axioms, and the business of the geometer is to reason intrinsically on this schema.

The process of constructing an abstract schema in correspondence with a given exterior significance may be called "abstraction by axiomatization."

Let us observe that the schema is the *abstract* of its exterior significance and that the latter is the *concrete* of the schema. Abstract and concrete are thus relative to one another. Their mutual correspondence, as well as their opposition, constitutes a part of their meaning.

E · Meaning of Intuition   There is, however, an important difference between the schematization of a city and that of geometry. We have no hesitation how to perform the first task. But it is not quite as clear how to go about picking for geometry "a system of statements having the value of axioms." When we considered the axiomatization of the straight line, we assumed that the notion of a straight line is familiar to us.

We know the thing "by intuition." Euclid's axioms have been accepted through the ages, as given "by intuition." Let us try to examine what this notion *by intuition* means.

Our accumulated knowledge is perpetuated and transmitted from generation to generation largely through books. More than 3000 years ago, a wise man voiced a complaint that "of making many books there is no end."[35] No part of this blame may be attributed to children, for children do not write books. But children, even infants, acquire a considerable amount of knowledge about the surroundings they live in and make a vast number of adaptations to it.

The adult, however, by the time he is ready to write a book, is prone to forget about the things he learned in his early life without the benefit of books. Relying on his personal recollections, any adult would, for instance, staunchly maintain that he had always been able to walk, if his observation made on his very youngest contemporaries did not shatter this quaint illusion.

Suppose you take two flat sticks, say, two rulers, of equal length, hold them in your two hands so that they cover each other, and then slide them part way one on the other. It would not occur to you to resort to measurements in order to settle the question: Which of the two uncovered parts is longer? You know for sure that those two parts are equal, and you have always known that to be true. Now, have you? No, there was a time when you did *not* know it. That was the time when, after having counted up five of your wooden toy blocks, you entertained high hopes that there might turn out to be six of them, if you arranged those same blocks in a different order.[36]

Knowledge of this sort is "intuitive" geometry. We accumulate a considerable body of this kind of information at an early age. By the time we are confronted with our first textbook on geometry, we are pleasantly surprized to find how much of the stuff we already know.[37] And by the time we feel called on to write on or about geometry ourselves, we pass

those things on for "common sense" and as "self-evident, intuitive truths" (Euclid), or for "knowledge a priori" (Kant),
while they are no more and no less than empirical information
that had been acquired very early in the hard and exacting
school of living and acting in a certain environment.

F · Foundations of Logic   One of the basic entities that
preoccupy the logician is the concept of "object." One commonly conceives of an object as a quantity of matter packed
into a portion of space and having well defined boundaries.
This is a very useful idea, well adapted to our human needs.
But when they are subjected to the closer scrutiny of the
physicist, these qualities of common objects, as well as others
that we associate with such objects, retain a validity that is
only approximate and provisional. We thus arrive at the conclusion that the conception of a "given object" is the outcome
of an effort of abstraction bearing on the shape, motion, and
other preceptive qualities of common objects. In other words,
the idea of an "object" derived from our everyday experience
is only summarily correct, just as is the idea of a straight line
that is suggested by a stretched string.

Following up this analogy, we may attempt to axiomatize
the concept of "any object" in order to facilitate its study from
the point of view of logic.

The process of systematic axiomatization will require a
further schematization of the idea of "object" itself, as well
as of the conjoint idea of the presence of the object somewhere,
or of its complete absence.

In the first place, the idea of "object" will have to be
divested of the notion of the object being present at a definite
spot or place. The idea of object will just remind us that the
object is, or is not. This idea of presence or absence, when
pushed farther, results in the idea of pure existence or non-
existence.

Furthermore, the strictly practical notion of the permanence of the object with regard to its own properties leads to the abstract notion of its pure identity with itself. This notion, combined with the notion of pure existence or non-existence, results in the abstract of existential identity.

To render this idea of existential identity more concrete, let us consider two different marks, say A and B. Suppose it so happens that neither could be written down without the other likewise being written, and that neither could be crossed out or erased without the same happening to the other. Moreover, the only matter of concern is to ascertain their presence or their absence. Under such circumstances, no confusion could arise if either one of the two marks A, B, is taken for the other. This practical equivalence of A and B is a good realization of their purely *existential equivalence*.

Going a step further, the same thing may be restated if, instead of two different marks, the same mark were drawn twice—that is, two distinguishable realizations of the same mark, say A, but again provided that only the question of presence or absence is involved. The two realizations may then be considered as identical. This practical identity realizes the abstract idea of *existential* identity.

Let us take the idea *any object* as our undefined term. We are now ready to formulate the axioms governing that term. It should be recalled that the axioms aim to point out, or reproduce, in connection with the abstraction considered, the salient properties of the concrete representations of that abstraction, or better, the properties that are inherent in the intuitive image we have of those concrete representations.

The realization of the abstract notion of existential identity by the practical equivalence of two copies of the same symbol quite naturally suggests the following.

Axiom 1. Any object is identical with itself.

To this axiom we shall add two more, based our practical experience and on our everyday knowledge of the rules of presence and absence of material objects:

Axiom 2. Any object is, or is not.

Axiom 3. No object can be and not be at the same time.

The three axioms represent, respectively, the principle of identity, the principle of the excluded middle, and the principle of contradiction.

The upshot of our efforts at abstraction by axiomatization is thus the idea of "any object" governed by the laws of being, or not being, and the law of existential identity with itself, but which is otherwise undetermined. This eminently abstract idea of an object might be called the "abstract object" or the the "logical object."

The concrete object realizes the logical object in the same way that the stretched string realizes the geometrical straight line. The idea of pure existence of the logical object is realized by a natural or concrete object in the same fashion and to the same degree as the ideal rectitude of a straight line is realized by the crude rectitude of the string. But no concrete object realizes the abstract object any more closely than it can realize the geometrical idea of a point.

These observations simply emphasize the fact that in the present case, as in any other abstraction by axiomatization, the relationship between the concrete and the abstract is adequate only in a schematic way. The degree to which it is necessary to simplify the common, intuitive notion in order to achieve this correspondence is clearly shown by our effort to establish the principle of identity.

G · Symbolic Representation, or Miniature Realization   Let A be the symbol of an object whose existence or nonexistence has not been specified, and let $\underline{A}$ and $\overline{A}$ be the symbols for A's existence and nonexistence, respectively. The three marks $\underline{A}$, A, $\overline{A}$ are three concrete objects. With reference to them, our three fundamental axioms take on the following forms:

Axiom 1. The letter A has everywhere the same significance.

Axiom 2. Each determination of A is expressed by its being either underlined or overlined.

Axiom 3. The latter two cases are mutually exclusive.

It is essential to observe that in those three statements all the words used have their ordinary meanings. There is no mention of purely existential identity or of pure existence. The reason for it is that in formulating these statements we have entrusted ourselves to our intuitive and practically certain knowledge concerning three signs drawn on paper.

To put it in other words, these three symbols and the formal rules that they have to obey are a miniature realization of the abstract schema that we have devised. This abstract schema is the link that establishes a correspondence between the three symbols A, $\underline{A}$, $\overline{A}$, the concrete number 2, so to speak, and the concrete number 1—namely, the common objects with which we started in the first place. However, we are prone to forget about the existence of this connecting link and see only the two concretes facing each other; the original concrete upon which we were reluctant to operate and the new concrete, much reduced and more readily handled.

The undetermined object symbolized by the "abstract form" A may be filled, or it may be empty. The concrete realization, or model of this form, may be perceived in any object that may, at will, be brought into the field of attention, or may be far from it. The form $\underline{A}$, on the other hand, symbolizes an existing object.

Now consider two objects that have no apparent tie and that may be treated independently of one another. True, a sufficiently close examination of any two objects is likely to result in the discovery of some kind of a relationship between them. But we shall gloss this over. If our two objects are, for instance, two books, their independence, for our purpose, may manifest itself in the fact that the two books may be in, or out of, the library independently of one another. This schematic form of independence is an abstract concept that may be adequately represented by two "forms of objects" admitting, without preference or distinction, the four eventualities shown in Table 1.

Table 1. Four eventualities.

No.	In words	In Symbols
1	A is and B is	$\underline{A}$ and $\underline{B}$
2	A is, but B is not	$\underline{A}$ and $\overline{B}$
3	B is, but A is not	$\overline{A}$ and $\underline{B}$
4	Neither A is nor B is	$\overline{A}$ and $\overline{B}$

The enumeration of these eventualities may perhaps incline the reader to think that any two objects are always independent. But that is not the case. The two independent objects that are existentially equivalent, considered in a previous paragraph, are just objects for which the eventualities 2) and 3) do not exist. We could also conceive of two objects for which the eventualities 1) and 4) are excluded, by definition. This is the case of mutual exclusion.

The case when only eventualities 1 and 4 are valid may be rendered concrete by two persons who always enter and leave a certain room at the same time; and the case when only eventualities 2 and 3 are valid is made concrete if one of those two persons leaves the room whenever the other enters, and vice versa.

These two examples show, by the way, that the relationships of equivalence and exclusion that we imagined between two forms are schematizing some close relationships that may exist between material objects.

H · Rational Theory of Objective Existence The concept of the set of the eventualities 1, 2, 3, and 4 considered as freely admissible may be referred to as the "abstract form relative to two abstract objects." If one, or two of these eventualities is left out, a less extended form, or a "subform," is obtained. Thus a pure and simple eventuality is the least extended form. We may also say that a subform "enters" into a more extended form, or that the latter "contains" the former, if all the eventualities of the former are also eventualities of the latter. Every subform enters into the complete form. Two subforms overlap, or are mutually exclusive, according to whether they do or do not possess a common eventuality.

Using this terminology, we may state the following new axiom: Two determined objects which enter into the form of equivalence cannot enter into the form of mutual exclusion. The realization (involving two persons) that we have considered makes it clear that this statement (axiom) formulates an empirical law of the world of material objects, a very primitive law, and therefore one of practically unfailing validity.

Thus, starting with the most common and the most elementary properties of material objects, and applying the axiomatic method as exemplified in plane geometry, we arrived at the concepts and the rules of pure existence. These ideas are basic in Gonseth's *Rational Theory of Objective Existence;* Gonseth's theory axiomatizes one of the first chapters of physics, if not the very first—namely, the one dealing with the existence, the presence, and the absence of objects of any kind. In other words, the physics of any object and the rational theory of pure existence are two phases of the same undertaking, the former being the external significance of the latter.

For the subject at hand, the importance of the rational theory of existence is that all the laws of elementary logic may be expressed in the form of rules of existence, as was quite apparent when we sketched those ideas, and may be substantiated by further analysis. But we shall not pursue this argument. The detailed treatment may be found in Gonseth's own writings.

I · Logic The conclusions to be drawn from the preceding discussion are as illuminating as they are far-reaching. The rules of pure existence being the schematized properties of common objects, the same holds for the equivalent laws of logic. Hence, the common sense of logic and its intuitive rules are seen to be the outcome of a schematization that is based upon our experience in the world of common objects.

Furthermore, since the abstract laws that logic formulates have their origin and their realization in the domain of concrete objects, those laws take on the significance of very primitive natural laws and are therefore practically infallible. That is what accounts for their usefulness, on the one hand, and for their irresistible power over us, on the other hand.

Our distinguished contemporary Maurice Fréchet (b. 1878), professor of mathematics at the Sorbonne, put the question of the origin of logic in a nutshell: "The rules of logic start with an approximation of the real, and that reality is rediscoverable even in the remotest conclusions drawn from those laws. Is it just by lucky accident, independently of all experience, that those laws impose themselves upon our mind? Or is not our acceptance of those laws from our predecessors due to the fact, taught us by our daily experience, that if we apply those laws correctly, we are never mistaken? We are thus not far from concluding that logic itself is a product of our experience, that logic is the result of an inductive synthesis. It is therefore quite legitimate, and even very useful, to submit logic to a process of axiomatization. This axiomatization, like that of any other science, must be considered as being only an essentially revisable schematization of the practical rules of reasoning. But we are certain that we shall always be able to utilize our logic, without change, in the major part of our scientific research." In brief, the empirical origin of logic is obvious a priori, despite the firm conviction of all those thinkers through the ages for whom the laws of logic were inherent laws of the mind. But then, the principle of duality turns out to be obvious, and so does Goedel's theorem. It would seem that nothing is more effectively hidden in the farthest recesses of obscurity than the obvious.

Since our laws of logic are derived from our practical experience, our reasoning can be valid only as long as we apply it to our environment as it is here and now, so to speak. This may cause trouble in some unsuspected and unsuspecting quarters. Take the great phalanx of enthusiastic space travelers, young and otherwise, that has sprung up in the wake of the rockets of recent invention. These travelers are beset by a great many worries and difficulties. But the fun-seeking excursionists to neighboring planets, as well as the intrepid conquistadores of new galaxies, may discover, to their amazement and chagrin, that the logic which they had found so reliable as long as they stayed home, goes "haywire" when they get abroad.

But remaining peacefully at home offers no guaranty of the permanent validity of our logic. Should our environment change, we would have to change our logic accordingly. This may sound fantastic, but it is not outside the realm of the possible. In fact, in a way, we are already in the midst of such a change right now, and have been for about half a century. Quantum theory, the new atomic theory, and the theory of relativity have confronted us with phenomena that operate on a scale either too vast or too minute compared with those on which our senses received their education and training. No wonder that we run into "inconsistencies" and "contradictions." The physicists have had to reexamine many of our notions that were well established according to our "common sense."[38] The logicians, for their part, try to meet the newly arisen problems by introducing multivalued logic.[39]

Our discussion of the origin of the laws of logic has brought into the open the limitations of those laws and should thus contribute to a better understanding of that wonderful instrument our human race is so proud of—the power of reasoning. We have convinced ourselves once more that our source of knowledge lies in closer contact with our environment. Is it not this idea that the Greek mythology wanted to express by imagining the demigod Anteus whose power endured as long as he maintained contact with the earth?

The same idea may be found in the play *Chantecler* of the French poet Edmond Rostand (1868-1918). This is the way the mighty Chantecler explains to his friend the pheasant hen where he derives his power to call out the sun from below the horizon: "I never start to sing until my eight claws, after clearing a space of weeds and stones, have found the soft, dark turf underneath. Then placed in direct contact with the good earth, I sing." We, too, have to be "in contact with the earth" if we want the light of knowledge to shine on us.

FOOTNOTES

[1] See Chapter I, Section 2c; Chapter III, Section 2.
[2] Bulletin American Mathematical Society, 1912-1913, p. 332.
[3] See Chapter I, Section 3f, section 4b; Chapter V, Section 1i.
[4] See Chapter I, Section 4c.
[5] For K. Goedel's contribution to the question of consistency of the postulates see Chapter I, Section 4b.
[6] Cf. Chapter I, Section 2a.
[7] *Enseignement Mathématique*, 1931, p. 29.
[8] See Chapter IV, Section 2d.
[9] Chapter I, Section 1f.
[10] See Chapter I, Section 4a.
[11] Cf. Chapter I, Section 4a; Chapter III, Section 2b.
[12] See Chapter I, Section 4d.
[13] *Mathematical Philosophy*, New York, 1922, pp. 412-413.
[14] See Chapter VII, Section 1c.
[15] Education Times, Reprints, V. 65, 1896, Question 12854.
[16] Cf. Chapter III, Section 3i.
[17] See Chapter V, Section 1e.
[18] Cf. Chapter III, Section 3f; Chapter V, Section 1j.
[19] See Chapter I, Section 1d, Section 4b; Chapter V, Section 1i.
[20] *Encyclopédie Française*, Vol. I, p. 20-26, 1937.

[21] *An Essay on the Psychology of Invention in the Mathematical Field*, Princeton University Press, 1945.

[22] Cf. Chapter V, Section 1i.

[23] Cf. Chapter I, Section 2d; Chapter III, Section 2.

[24] Cf. Chapter V, Section 1c.

[25] See Chapter I, Section 1d, 3f; Chapter V, section 1i.

[26] *Foundations of Science*, H. Poincaré, Edited by G. B. Halstead, Science Press, Lancaster, Pa., 1913, 1946.

[27] Cf. Chapter I, Section 1d; Chapter V, Section 1i.

[28] Chapter I, Section 1e.

[29] *Le décline des absolus mathématico-logiques*, G. Bouligand and J. Desgranges, Sedes, Paris, 1949, p. 16.

[30] *The Axiomatic Method in Biology*, J. H. Woodger, Cambridge University Press, Cambridge, 1937.

[31] "Vision" in *Insect Physiology*, V. G. Dethier, Edited by K. D. Roder, Wiley, New York, 1953, p. 488.

[32] *Radio Astronomy*, B. Lovell and J. A. Clegg, Wiley, New York, 1952. *Scientific Monthly*, W. L. Roberts, 79, 170, 1954. *Astrophysical Journal*, articles by Minkowski and others, 1954.

[33] *Problems of Science*, F. Enriques, Open Court, Chicago, 1914.

[34] Cf. Chapter I, Section 4d.

[35] Ecclesiastes: 12, 12.

[36] *Pedagogical Seminary*, 27, 75, "Number, time and space in the first five years of a child's life," S. R. A. Court, 1920.

[37] Cf. Chapter I, Section 2a.

[38] *Scientific Monthly*, 79, '2, P. W. Bridgeman, 1954.

[39] See Chapter V, Section 1l.

Chapter II

SOME SOCIOLOGIC ASPECTS
OF MATHEMATICS

1 · MATHEMATICS AND CIVILIZATION

A · The Early Beginnings of Counting and Reckoning In a country where compulsory school attendance has been the practice for several generations the vast majority of children learn to count, that is, to recite the series of words one, two, three, four, and so on, very early, long before they reach school age. One of the consequences of this situation is that when they grow up they cannot remember when they could not perform this very useful trick.

The same situation prevails with regard to the human race as a whole. We know quite definitely that there was a time when the notion of number was totally alien to mankind. Who was the genius who first asked the momentous question: "How many?" We will never know. At a certain stage of social development the need arises to determine how many objects constitute a given collection. The answer to the question becomes a social necessity. Contributions toward finding that answer are made by individuals confronted with the same need, and the notion of number slowly emerges.

How slow and painful a process of creation this was may be judged from the fact that there are human tribes whose languages have no words for numbers greater than four, and even no greater than two. Beyond that any group consists of "many" objects.

A bright light is shed upon this subject by a story told the writer by a colleague, from the latter's personal experience, about a flock of crows.

The birds were infesting a cornfield. One morning, when two men armed with shotguns approached the field, the crows took refuge in a grove of trees at one end of the field. They remained there as long as the two suspected enemies occupied a shed at the other end of the field. When one of the men emerged from the hiding place and left the scene, the birds were not impressed: they remained where they were. But when the other man left the shed and vanished in the distance, the hungry birds resumed their feasting.

Next day three men entered the shed and the flock perched in the trees. The cautious birds were not fooled when two of the men came out of the shed and walked away. They waited until the third one did likewise. The following day three of the four men in the shed came out of the hiding place. As soon as they got out of sight, the crows descended upon the field in force. The wise crows could apparently tell the difference between one and two, also between two and three, and acted accordingly. But the difference between three and four passed the limits of their arithmetical wisdom, and the flock paid a high tribute for their ignorance.

Our numbers are applied to any kind of object in the same way, without discrimination. They have a kind of "impersonality," which was not the case with primitive man. With him the number applied to a group is modified in accordance with the nature of the group. The number characterizes the group in the same way as an adjective applied to a noun modifies the object to which it is applied. The English language has preserved some traces of that attitude. A group of cattle is a *herd,* while a group of birds is a *flock;* a group of wolves is a *pack,* while a group of fish form a *school.* It would be shocking indeed to speak of a school of cows. Other languages offer much more striking proofs of such an attitude towards numbers in their relation towards the objects they are applied to. Thus in English we use the singular grammatical form when one object is involved, and we use the plural grammatical form for any number of objects larger than one. Some of the languages of the Western world, in their earlier stages of development, had a special grammatical form, a dual form, when two objects were spoken of. Some languages even had separate grammatical forms when reference was made to three objects, and still another for four objects. An instructive example of the way the form of the same number may be modified to fit the group to which it is applied is furnished by the Polish language in its use of the number two. In that language a different form of "two" is used when applied to two men, to two women, to a man and a woman, and to inanimate objects or animals. These forms are, respectively: *dwaj, dwie, dwoje, dwa.*

The process of accumulating enough words to answer the question: how many? to satisfy the growing needs was slow and laborious. Man derived a great deal of help from the natural set of counters he always carried with him—his fingers.[1] We still use the word "digit" both for fingers and to designate a number less than ten. Of the many examples that could be cited to illustrate the use of fingers as counters let us quote a report of Father Gilij who describes the arithmetic of the Indian tribe of the Tamanacas, on the Orinoco river.

The Tamanacas have words for the first four numbers. When they come to five they express it by a phrase which literally means "a whole hand;" the phrase for the number six means literally "one on the other hand," and similarly for seven, eight and nine. When they come to ten they use the phrase "both hands." To say eleven they stretch out both hands, and adding a foot, they say "one on the foot," and so on, up to 15, which is "a whole foot." The number 16 is "one on the other foot." For twenty they say "one Indian," and 21 is expressed by saying "one on the hands of the other Indian;" forty is "two Indians," sixty "three Indians," and so on.

In this connection it may be of interest to point out that the Russian word for five ("piat") is a slight modification of the word for fist ("piast"). The same is true for other Slavic languages.

When the question: how many? has once been raised, mere counting becomes insufficient. Further steps in civilization bring about the need of computation. The strongest single factor that stimulated the development of methods of computation was trade. According to the mythology of the ancient Egyptians, arithmetic was invented by their God of commerce. As with counting, the beginnings of reckoning were slow and laborious, awkward and painful. A trader in tropical South Africa during the last century has this to say about the members of the Dammara tribe. "When bartering is going on, each sheep must be paid for separately. Thus, suppose two sticks of tobacco to be the rate of exchange for one sheep; it would sorely puzzle a Dammara to take two sheep and give him four sticks. I have done so, and seen a man put two of the sticks apart and take a sight over them at one of the sheep he was about to sell. Having satisfied himself that *that* one was honestly paid for, and finding to his surprise that exactly two sticks remained in his hand to settle the account for the other sheep, he would be afflicted with doubt; the transaction seemed to come out too "pat" to be correct, and he would refer back to the first couple of sticks; and then his mind got hazy and confused, and he wandered from one sheep to the other, and he broke off the transaction, until two sticks were put in his hand, and one sheep driven away, and then two other sticks given him and the second sheep driven away." It would seem that at least to this representative of humanity it was not obvious that two times two makes four.

The story illustrates the blundering beginnings of the art of reckoning. To relate the evolution of this art from its humble beginnings to the heights of power and perfection it has achieved in modern times, and how this art has followed and served the ever growing needs of mankind is to tell one of the most exciting sagas in the history of civilization. Only a mere outline can be attempted here.

Various human activities, and in particular commerce, require the keeping of some numerical records. Some kind of marks had to be invented for the purpose. The devices used through the ages were knots tied in a rope and notches cut in sticks. It may surprise some readers that such sticks, called *tallies*, were used as a method of bookkeeping by the Bank of England well into the nineteenth century.

The first written symbols for numbers were, naturally, sticks: *One stick, two sticks, three sticks,* and so on, to represent "one," "two," "three" etc. This worked fairly well as long as the numbers to be represented were small. For larger numbers the sticks occupy too much space, it becomes difficult to count them, and it takes too much time. The sticks had to be condensed into groups, thus representing larger units, and these new units in turn had to be condensed into larger units and thus a hierarchy of units had to be formed.

This need for condensation of numerical symbols is readily brought home to us by a familiar example. In theory the treasury of the United States should mint only one kind of coin, namely a penny, for every sum of money can be realized with pennies. In practice, however, this would be a most awkward procedure, even when only small sums are involved. To help matters the treasury mints also nickels and several kinds of silver coins. Moreover, the treasury considers that one is justified in refusing to accept more than twenty five pennies in any single payment. For larger amounts the treasury condenses one hundred pennies into a single paper dollar bill, and then continues the process by issuing bills of several higher denominations.

The Greeks and the Hebrews used the letters of their alphabets as numerals. The Babylonians had special numerical symbols. The Roman numerals are still in use occasionally, as for instance on our clocks. All these symbols or marks for numbers had one feature in common—they did not lend themselves to arithmetical computations. The art of reckoning had to be carried out with the help of different devices, the chief among them being the counting frame, or the *abacus*. This instrument most often consisted of a rectangular frame with bars parallel to one side. The operations were performed on the beads or counters strong on these bars. This instrument was widespread both in Asia and in Europe. When the Europeans arrived in America they found that a form of abacus was in use both in Mexico and in Peru.[2]

The method of writing numbers and computing with them that we use now had its origin in India. The most original feature of that system, namely the zero, the symbol for nothing, was known in Babylon and became common in India during the early centuries of the Christian Era. This system of computation was brought to Europe by the Arabic and Jewish merchants during the twelfth century. The first printing presses set up in Europe, in the middle of the fifteenth century, rolled off a considerable number of commercial arithmetics. Two centuries later the abacus in Western Europe was little more than a relic of the past. It is still widely and efficiently used in the Orient.

The very heavy demands that modern life in its various phases makes upon computation seem to be turning the tide against paper and pencil reckoning. We are about to enthrone the abacus back again, in a much improved form, to be sure, but nevertheless in the form of an instrument. In fact, we are using a considerable number of them, like the slide rule, the cash register, the various electrically operated computers, to say nothing of the computing machines which operate on a much higher level, like those which give the solutions of differential equations. Such is the devious and puzzling road of human progress.

B · Measuring. Beginnings of Geometry and Chronology
"How many?" This question is the origin of arithmetic and is responsible for much of its progress. But this question cannot claim all the credit. It must share the credit with another, a later arrival on the scene of civilization, but which is even more far reaching. This question is: "how much?" How much does this rock weigh? How much time has passed between two given events? How long is the road from town A to town B? etc. The answers to these questions are numbers, like the answer to the question: "how many?" There is, however, a vast difference between the numbers which answer the two kinds of questions.

The answer to the question: "how many?" is obtained by counting discreet objects, like sheep, trees, stars, warriors, etc. Each of the objects counted is entirely separate from the others. These objects can be "stood up and be counted." Something vastly different is involved in the question: How much does this rock weigh? The answer can only be given by comparing the weight of the given rock to the weight of another rock, or to the weight of some other object taken for the unit of weight, say a pound or a ton. Obviously this is a much more involved process and implies a much more advanced social and intellectual level than the answer to the question: how many?

The question: "how many?" is always answered by an integer. Not so the question: "how much?" Given 17 trees, is it possible to plant them in five rows so that each row has the same number of trees? The answer is: "No," and this is the end of the story. But given seventeen pounds of salt in a container, it is possible to distribute this salt into five containers so that each of them will hold the same amount of salt. But the question: "How many pounds of salt does each container hold?" cannot be answered by an integer. Thus, the question: "how much?" is responsible for the invention of fractions. It is also responsible for the introduction of irrational numbers. But about that we may say something later on.

The question: "how much?" that is, the introduction of measurements, has involved us in another kind of difficulty which did not bother us in connection with the question: "how many?" We can ascertain that the group at the picnic consisted of forty boys. But when we say that this table is forty inches long, we can only mean that it is closer to forty inches than it is either to 39 or 41 inches. We may, of course, use more precise instruments of measurement which may narrow down the doubtful area, but it will not remove it. Results of measurements are necessarily only approximations. The degree of approximation to which we carry out these measurements depends upon the use we are to make of these measured things.

The herdsman is much concerned with the question: "how many?" The shepherd, in addition, is also interested in the question: "how much?" after he is through sheering his flock. When a human tribe turns to agriculture, the question: "how much?" imposes itself with increased insistence. Agriculture requires some methods of measuring land, of measuring the size of the crop, that is measuring areas and volumes as our school books call it. Furthermore, the agricultural stage of society implies already a considerable degree of social organization, and the tax collector appears on the scene. This official is vitally interested in the size of the crop. He also has to have some numerical records of the amount of taxes collected and of the amount of taxes due. Now you many not like the tax collector. Few people waste too much love on this maligned official. It is nevertheless quite obvious that no organized society is possible without the collection of taxes, that is without contributions from the individual members of that society towards the necessary enterprises that are of benefit to the members of the entire community. And such collections cannot be made in any orderly fashion, unless answers can be given to the two questions: "how much?" and "how many?"

How far back in the history of mankind the question "how much?" was first asked we can only guess, and that very roughly. These surmises are helped by the study of the culture of some of the primitive tribes still inhabiting this earth, or did so in recent past.

However, conjectures are replaced by documentary evidence when we turn to the period of human history which starts about six or seven thousand years ago in Mesopotamia, Egypt, India. This is the period of The Bronze age, the beginning of urban civilization. The Babylonian tablets, the Egyptian papyri, and other documents tell us a new and wonderous story of new forms of social and governmental organizations, of remarkable conquests in the domain of arts and crafts, of great expansion of trade and commerce. The same documents tell us of astonishing achievements in the field of mathematics and astronomy. The historian of civilization makes it clear that this new knowledge was called forth—and contributed to —by the artisan, the builder, the merchant, the surveyor, the warrior.[3]

The cultivation of the land faced the human race with problems of geometry. Egypt with its peculiar dependence upon the flood waters of the river Nile was confronted with extra difficulties of a geometrical nature. That is the reason why geometry found such a fertile soil in the valley of the Nile.

Much geometry had to be discovered in order to construct human habitations. When civilization progresses beyond the cave dwelling stage, shelter becomes a problem of the first magnitude. The construction of dwellings involves in the first place knowledge of the vertical direction, as given by the plumb line. It was observed very early that the plumb line or a pole having the same direction as the plumb line makes equal angles with all the lines passing through its foot and drawn on level ground. We have thus what we call a right angle, as well as the famous theorem of our text books that all right angles are equal.

However important the answers to the question: "How much?" may have been in the connections we just considered, the most important answer to this question is the one connected with the measuring of time. With the most rudimentary attempts at agricultural activity comes the realization that success is dependent upon the seasons; this dependence is even exaggerated. We still worry about the phases of the moon when we want to plant our potatoes.

Various tribes on the surface of the globe noticed that the shortest shadow cast by a vertical pole during the day always has the same direction. This is the north and south direction. The sun at that time occupies the highest point in the sky. It is essential to have a way of marking this direction. Here is how it can be done.

A circle is drawn on the ground having for center the foot of the pole used in the observation. The two positions of the shadow are marked, the tips of which just fall on the circumference. The north-south line sought is the line mid-way between the two lines marked, and that north-south line was found by many human tribes by bisecting this angle, and was done by the methods still in use in our text-books.

Measurements connected with the sun, the moon, and the stars in general cannot be made directly. Some round-about method must be used. Neither could the size of the earth be determined directly. On the elementary level such artifices are based on geometry and trigonometry. Two centuries B. C. Eratosthenes, the librarian of the famous Alexandrian Library, succeeded by the use of such methods in determining the length of the diameter of the earth with a surprising degree of accuracy. He thus made his contemporaries realize that the world they knew was only a very small part of the surface of the earth.

These sketchy indications give an idea of the role mathematics played in the development of mankind from the earliest times up until the great civilizations of antiquity.

C · The Renaissance Period. The Great Voyages. The Invention of Analytic Geometry and of the Calculus   The Renaissance was the age of the revival of secular learning in Europe. It was also the age of the great voyages and of the discovery of America, the age of gun-powder and of mechanical clocks.

The new interest in seafaring had raised many pressing problems that had to be solved. The most obvious one was the need for a way of determining the position of a ship on the high seas, that is the need of determining the longitude and the latitude of the ship at any time. This involved a great deal of laborious computation. The invention of logarithms reduced this labor to a fraction of the work it used to require. This accounts for the great success that the invention of logarithms enjoyed, as soon as it became available.

The process of finding the longitude required an accurate clock which could be relied upon. We mentioned before the important role the need of determining the seasons played in history of civilization, and the mathematical problems that had to be solved in this connection. The navigation of the Renaissance required the measuring of time with great precision. It was a question not of seasons and days but of minutes and seconds. The instrument that made such accuracy possible was the mechanical clock moved by a pendulum, then by springs. This moving mechanism raised many problems of a mathematical nature that the available mathematical resources were insufficient to cope with. New mathematical methods were needed.

New mathematical problems were also raised by the cannon. It may be observed, in passing, that a cannon was just as much a necessary piece of equipment of a ship starting out on a long voyage towards unexplored shores as was a map, or a clock.

A gunner frequently needs to determine the distance to certain inaccessible objects. The information has thus to be obtained by indirect measurements. This is a problem that was met with much earlier in the history of civilization and was solved in various ways. The cannon has stimulated further development in this connection, thus contributing to the progress of trigonometry.

But artillery presented problems of a new type. The cannon ball was an object which moved with a speed that was unprecedented in the experience of man. Motion took on a new significance and called for mathematical treatment and study. It required the study of the path that the projectile describes in the air, the distance it travels, the height it reaches at any given distance from the starting point, and so on. In short it required what we now call a graph.[4]

The computation of the longitude of a ship at sea is based on astronomical observations and computations made in advance and published for that purpose. The greater the accuracy of these data, the more correctly can the position of the ship be determined. Thus navigation made necessary a more accurate knowledge of the motion of heavenly bodies.

The mathematics that the Renaissance inherited from preceding periods was inadequate for the study of motion. The new mathematical tools that were invented for the purpose of answering the new questions raised were: (1) Analytic Geometry, invented by René Descartes (1637)[5] and (2) the Infinitesimal Calculus, the contribution of Newton and Leibniz to the learning and technical proficiency of man.[6]

The path of a cannon-ball, or, for that matter, the motion of any object is most readily studied by a graphical presentation of that motion. Nowadays graphs are very common. We see them even in the newspapers when things like the fluctuation of the price, say, of wheat is discussed. But it took nothing less than the invention of Analytic Geometry to put this simple and powerful device at the service of man.

If a body travels along a curved path, it does so under the action of a force exerted upon it. If the force suddenly stops, the moving object continues nevertheless to move, not along the curve, however, but along the tangent to that curve at the point where the object was when the action of the force ceased. Thus, in the study of motion, it is important to be able to determine the tangent to the path at any point of that curve. The resources that mathematics had to offer up to the middle of the seventeenth century were insufficient to solve that apparently simple problem. The differential calculus provided the answer.[7]

The calculus provides the tools necessary to cope with the questions involving the velocity of moving bodies and their acceleration, or pick-up. The ancients had only very hazy notions about these concepts. The unaided imagination seems to find it very difficult to handle them successfully. The methods furnished by the calculus take all the sting and all the bitterness out of them. When velocity and acceleration are presented to students of mechanics who do not have the calculus at their disposal, these notions are still explained in terms of the calculus, in a round-about, disguised fashion.

D · Mathematics for the Modern Age   Our own age is confronted with technological problems of great difficulty. The mathematical tool they call for were not in existence at the time of Newton, two centuries ago. The airplane alone is sufficient to make one think what a variety of questions of an unprecedented kind had to be answered, what complicated problems had to be solved to enable the flier to accomplish all the wonders of which we are the surprised and admiring witnesses. The difficulties of constructing the airplane wings necessitated the concentration of mathematical talent, and mathematical information that has hardly any parallel in history.

As has been pointed out, sea-faring called for the solution of many problems. However, a ship sailing the high seas has one important feature in common with a vehicle traveling on land: both move on a surface. From a geometric point of view the problems related to their motion are two-dimensional. An airplane that roams in the air above is engaged in three-dimensional navigation. The geometrical aspect of flight belongs to the domain of Solid Geometry, and the problems connected with it are thus much more difficult, other things being equal.

Mathematics plays an enormous role in the field of social problems, through the use of statistics. I have already pointed out the value of mathematics in connection with the collecting of taxes, at earlier stages of civilization. The functions of a modern government are vastly more complex, more varied, and applied on an enormous scale. The variety and scope of problems modern government is interested in can be gleaned from the questions the citizen is asked when he receives the census blank, every ten years. To study the wealth of information that is thus gathered on millions of blanks is the function and the task of the census bureau. The inferences that can be drawn from these data are as involved as they are far-reaching in their applications. Such a statistical study requires a wide range of mathematical equipment, from the most elementary arithmetic to the most abstruse branches of mathematical analysis. If one thinks of the new functions of social welfare that the government has taken on, like social security or old age pensions, as well as of those that are in the offing, like health insurance, and the millions of individuals that these services cover, one is readily led to the realization that the intelligent dealing with these services sets before the government new statistical problems of vast magnitude.[8]

The government is not the only social agency to use statistics. Far from it. Insurance companies have been using statistics for a long time. Banks and other organizations which study the trends of business arrive at their predictions by statistical analysis. The study of the weather raises many very difficult statistical problems. Statistics is used to determine the efficiency of the methods of instruction in our public schools. This list could be made much longer and become boring by its monotony. As it is it will suffice to convey the idea of the all-pervading role this branch of applied mathematics plays in our modern life.

E · Conclusion   We have alluded several times to the fact that during the course of the centuries mathematics was called upon to provide solutions for problems that have arisen in various human pursuits, for which no solution was known at the time. This, however, is not always the way things occur. In many cases the reverse is true. When the need arises and the question is asked, mathematics reaches out into its vast store of knowledge accumulated through the centuries and produces the answer. The astronomer Kepler had before him a vast number of observations concerning the motion of the planets. These figures were meaningless until he noticed that they would hang nicely together if the planets followed a path of the form which the Alexandrian Greek Apollonius called an ellipse. Another plaything of the same Apollonius, the hyperbola, came in very handy to locate enemy guns during World War I, when the flash of the gun could be observed twice.

This readiness of mathematics goes much further. Various branches of science, when they pass and outgrow the purely descriptive stage and are ready to enter the following, the quantitative stage, discover that the mathematical problems which these new studies present have already been solved and are ready for use. Thus Biology has in the last decades raised many questions, answers for which were available in the storeroom of mathematics. At present the scope of mathematics used in "Mathematical Biology" exceeds by far the mathematical education which our engineering schools equip their graduates with. A similar tale can be told of psychology, economics, and other sciences.

We have tried to point out the close relation of the mathematics of any period of civilization to the social and economic needs of that period. Mathematics is a tool in the work-a-day life of mankind. It is closely connected with the well-being of the race and has played an important role in the slow and painful march of mankind from savagery to civilization. Mathematics is proud of the material help it has rendered the human race, for the satisfaction of these needs is the first and indispensable step that must be taken before higher and nobler pursuits can be cultivated.

The Russian fable writer Ivan Krylov observed: "Who cares to sing on a hungry stomach?" The Hebrew sages of yore put it more concretely: "Without bread there is no learning," and they are not slow to follow it up with the converse proposition: "Without learning there is no bread." Three millennia or so later both propositions found an eloquent defender in one of the greatest minds of all time. Henri Poincaré in his philosophical writings refers to a dispute between those of his contemporaries who thought that we should study mathematics in order to build machines, and their opponents who thought that we should build machines in order to have leisure to study mathematics. Poincaré opines that he is in complete and perfect agreement with both camps and endorses both propositions.

2 · MATHEMATICS AND GENIUS

A · The "Heroic" and the "Objective" Interpretations of History Once upon a time, many, many years ago, so the story goes, a beautiful stallion was brought to the royal court and presented to the king. The stallion was very wild. The king was warned that no man had ever managed to mount the fiery beast. The heir apparent who happened to witness the presentation ceremony of this unusual gift, jumped upon the back of the spirited horse, and before anybody had time to realize what was happening, the young prince was already way out of sight. The King's anxiety for the safety of his beloved son was very great. After a certain lapse of time, the young man reappeared, safe and sane, on the back of the subdued, tame animal. The proud and loving father was so elated that he exclaimed in exaltation: "My son, find for yourself another kingdom. Mine is too small for you." These accidental words of the king took deep root in the sensitive soul of the young prince. History knows this young man under the name of Alexander the Great (356-323 B. C.), the famous conqueror of the ancient world.

I read this story in my school-text on ancient history, a fine book, full of names and dates. Every historical event had its precise moment of occurrence recorded. You were told exactly by what king, or general, or by what great leader any given event was brought about. For the sake of brevity let us refer to this way of conceiving historical events as the "heroic view" of history.

This heroic interpretation of history is very attractive, because of its simplicity and its definiteness. All the whys and wherefores are readily answered by the names of the great men who made the history of the nation, or of the race. However, this heroic view has an obvious weakness: it makes history whimsical, capricious, and accidental, to the point of triviality. Suppose that our stallion of a moment ago, in its frantic effort to rid itself of its unsuccessful and unlucky tamers, had broken a leg, or two. King Phillip would have been deprived of the occasion to utter those fateful words of his, and his son Alexander would have lived out his life as an obscure and inconsequential ruler of the little kingdom of Macedonia.

According to a much repeated saying, of undetermined origin, "God made George Washington childless, so he could become the father of his country." Thus, if it were not for some physiological peculiarity or deficiency of Martha Washington (or was it of George himself?) this country would have remained a British colony, even unto this very day and generation.

During the nineteenth century various writers, like the Englishman Henry Thomas Buckle (1821-1862), the Frenchman Hippolyte Taine (1828-1893), best known in the English speaking world for his history of English literature, the German Karl Marx (1818-1883) have advanced the view that human history is not made by individuals, but is dominated by objective factors, line climate, geographic environment, natural resources, economic and social conditions, etc. This objective interpretation of history has since gained a great deal of ground. A forceful presentation of this conception may be found in the presidential address delivered before the American Historical Association by Edward P. Cheyney (1861-1947), under the title "Law in history"[9] in which the following two passages occur: "History, the great course of human affairs, has not been the result of voluntary action on the part of individuals or groups of individuals, much less of chance, but has been subject to Law." "Men have on the whole played the part assigned to them: they have not written the play. Powerful rulers and gifted leaders have seemed to choose their policies and carry them out, but their choice and success with which they have been able to impose their will upon their time have alike depended on condition over which they have had no control".

The heroic and the "objective" interpretation of history are obviously poles apart. Which of them is right? General human history is so many-sided, so complex, that it is easy enough to emphasize one element or another of its vast contents and arrive at conclusions which are contradictory, and still have each a good deal of truth in them. We may try to simplify the problem, as we often do in mathematics, reduce the number of variables, and examine a few of them at a time.

Our objective may perhaps be achieved more readily if we examine the history of a restricted, particular domain, say, that of mathematics.

B · Are Inventions Inevitable?   We are accustomed to pronounce with respect and admiration, not to say with reverence and awe, names like Euclid, Archimedes, Descartes, Newton, Leibniz, Lagrange, Gauss, Poncelet, Klein, Poincaré, and many others. We know the books those men have written, the theorems which bear their names. In our own time we know by name men who live in our midst and some of whom we know personally, men who lend luster and glory to our generation, men who give us courage and inspiration. Through the study, direct and indirect, of the works of these eminent scholars we know what they have contributed to the growth and advancement of mathemathical science. There hardly can be a more forceful confirmation of the importance of the individual in history, of the heroic interpretation of history, if you will. Nevertheless, there is another side to this medal.

On December 21, 1797, in Paris, the great mathematicians Laplace and Lagrange were both present at a brilliant social gathering which included a great many celebrities. Among the guests was also a victorious young general whose star was ascending rapidly, and who happened to be a former student of Laplace. In the course of the evening the general, while talking to the two world famed scholars, entertained them with some unusual and curious solutions of well known problems of elementary geometry, but solutions with which neither of his two eminent listeners were familiar. Laplace, a bit peeved, finally said to his erstwhile pupil, "General, we expect everything of you, except lessons in geometry". The name of the young general was Napoleon Bonaparte. Napoleon had learned about those strange constructions during his famous campaigns it Italy, whence he had just returned. While there, he met Lorenzo Mascheroni, a professor at the University of Pavia, who that very year, 1797, published a book *Geometria del Compasso* in which the author showed that all the constructions that can be carried out with ruler and compass, can also be carried out with compass alone, a very astonishing result, indeed. Had Mascheroni died in infancy, would science have been deprived forever of those Mascheronian constructions? One may think the question preposterous, for such a hypothetical query admits of no answer, one way or the other. Curiously enough, in the present case the question can be answered, in a very definite way. A century and a quarter before the publication of Mascheroni's book a Danish mathematician Georg Mohr published in Amsterdam a book in two languages, one in Danish and the other in Dutch, simultaneously, in which he gives Mascheroni's main result, as well as the solutions of a good many of the problems solved later by the Italian scholar. Mohr's book passed entirely unnoticed by his contemporaries. It came to light in the present century by accident. In the preface to his book Mascheroni states explicitly that he knows of no previous work along the same lines as his book, and there is not the slightest reason to doubt his word.

The story emphasizes the fact that so many mathematical discoveries, great and small, have been made independently by more than one scholar. This multiplicity of claims to the discovery of one and the same thing is probably the most outstanding fact in the history of mathematics.

The dispute as to whether Newton or Leibniz invented the calculus is well known.[10] The French claim, with a good deal of justice, that Fermat anticipated both of them. It is only Fermat's strange and persistent aversion to the pen that deprived him of the credit as inventor of that powerful mathematical tool.

A similar story may be related about the epoch-making discovery of analytic geometry. There is as much reason to refer to this discovery as "Fermatian" as there is to call it "Cartesian." Carl B. Boyer in the preface to his "History of Analytic Geometry"[11] says: "Had Descartes not lived, mathematical history probably would have been much the same, by virtue of Fermat's simultaneous discovery (of analytic geometry)."

The geometric interpretation of complex numbers was discovered independently and almost simultaneously by four different men, at the beginning of the nineteenth century. An instrument for drawing a straight line without the use of a ruler, known as the "cell of Peaucellier" (1832-1913), was also invented by a young student Lipkin of St. Petersburg (Leningrad).[12]

Even whole theories have grown up, the paternity of which nobody can claim with justice. A good and simple example of this kind is offered by the theory of inversion. This theory came into being early in the nineteenth century, and from so many different quarters that it is impossible to associate any particular name with it. The only thing that can be said about it is that, like Topsy, it "just growed."

The multiplicity of claims to the same discovery is so common that not only have we stopped to be surprised by it, but we have grown accustomed to expect it. Better to be able to protect the priority right of contributors, most of the editors of mathematical journals add to each article they publish, the date when that paper was received in the editorial office.

What was said about mathematics may be repeated with equal force about astronomy, physics, chemistry, mechanics, in fact about any science, pure or applied. Two industrious sociologists compiled a list of inventions, each of which has more than one claimant to its paternity. The list contains 148 entries and is far from being exhaustive. Armed with their incredible, but correct list, the two authors fire, point blank, an amazing question at their readers, namely: "Are inventions inevitable?"[13]

C · Genius and Environment   We are prone to think that the essence of genius is freedom. Does not genius invent or create what he will? On closer examination, however, it is seen that this conception of genius is an exaggeration. What a genius may accomplish depends upon circumstances which can be controlled by no individual. The invention of the creative individual is necessarily an extension of the knowledge of his time, or is something that satisfies the needs of his contemporaries. These characteristics have to be incorporated in the invention, if the genius is to be recognized as such. If a self-taught scholar from somewhere in the hinterland would send to the editor of a journal or to the Academy of Science a manuscript which in substance would amount to the discovery, say, of non-Euclidean geometry, or of the sextant, not much fuss would be made about the author, even if his honesty would not be called into question. And such things happen, on various levels of achievement. About the middle of the nineteenth century the Academy of St. Petersburg was offered by a teacher in some rural elementary school a crude exposition of the basic ideas of the calculus.

On the other hand, what genius can accomplish depends upon what others have done before. Newton realized that if he had seen farther than others, it is because he was "standing on the shoulders of giants." "Perhaps nowhere does one find a better example of the value of historical knowledge for mathematicians than in the case of Fermat, for it is safe to say that, had he not been intimately acquainted with the geometry of Appollonius and Viète, he would not have invented analytic geometry."[14] On the other hand, as great a genius as Archimedes could not invent analytic geometry, for the algebraic knowledge necessary for such an achievement was not available in his time.

The relation between the genius and the culture he is born into is expressed by A. L. Kroeber in the following way:[15] "Knowing the civilization of a land and of an age, we can then substantially affirm that its distinctive discoveries, in this or that field of activity, were not directly contingent upon the personality of the actual inventors that graced that period, but would have been made without them; and that, conversely, had the great illuminating minds of other centuries been born in the civilization referred to instead of their own, its first achievements would have fallen to their lot. Ericsson or Galvani, eight thousand years ago, would have polished or bored the first stone; and in turn, the hand and mind whose operations set in inception the neolithic age of human culture, if held in its infancy in unchanged catalepsy from that time until today, would now be devising wireless telephones and nitrogen extracts," or (let us add) nuclear weapons and interstellar ships, a generation or two later.

The dependence of the individual, whatever his natural endowments, upon the time and civilization he happens to live in, becomes quite obvious, once attention is called to this phenomenon. We are not a bit surprised to see that the French children are so very partial to the French language, and that the Chinese children, not to be outdone by the French, speak as unanimously the Chinese tongue. The same may be said, in a broader sense, about arts and crafts, music, or any other component element of culture. On a larger scale, analogous remarks may be made about those parts of culture which have become common to a considerable part of mankind, like the sciences, and mathematics in particular.

These observations may help us to comprehend the reasons for the multiplicity of claims for the same discovery. The anthropologist Leslie A. White puts it this way:[16] "In the body of mathematical culture there is action and reaction among the various elements. Concept reacts upon concept: ideas mix, fuse, form new syntheses. When this process of interaction and development reaches a certain point, new syntheses are formed of themselves. These are, to be sure, real events and have their location in time and space. The places are, of course, the brains of men. Since the cultural process has been going on rather uniformly over a wide area of population, the new synthesis takes place in a number of brains at once."

Tobias Dantzig (1884-1956) in his admirable book, *Number—The Language of Science*,[17] says the same thing, with a different emphasis: "It seems that the accumulated experience of the race at times reaches a stage when an outlet is imperative and it is merely a matter of chance whether it will fall to the lot of a single man, two men, or a throng of men to gather the rich harvest."

D · Genius and the "Instinct of Workmanship"   Granting that objective conditions determine the kind of discoveries that can be made at any given period of history on the one hand, and that on the other hand such inventions are "inevitable," such forward steps do not take place automatically. Each particular advance requires an effort, and often a very strenuous one, on the part of the gifted individual, the "genius" who brings it about. What impulse does the individual respond to, when he makes the requisite effort?

Mathematics, like any other science, in its early stages developed empirically for practical, utiltarian purposes. The demand for its services never cease through the ages, although the extent and the pressure may vary widely from one period to another, and mathematical inventiveness may vary accordingly. This is quite clear, for instance, in the case of the rapid strides made by mathematics during the brilliant seventeenth century.[18] The mathematician, like any other scientist, is not unmindful of the needs of his time and is not indifferent to the acclaim that would be his if he supplied the answer to a pressing question of his day.

There is, however, another phase of the situation to be considered. After a sufficient amount of mathematical knowledge has been accumulated, the cultivation of this domain of learning may become an interest in itself. Those versed in its secrets and adept in manipulating them may find it attractive to strive for new results just to satisfy what Thorstein Veblen (1857-1929) called the "instinct of workmanship."[19] The only extraneous element in the case may be the wish to gain the approval of the restricted audience of likeminded people or perhaps to confound some rivals. Outside of that the reward that may accrue to the mathematician for his efforts is to live through the pains of creation and to experience the exhilarating joy of discovery. His is a labor of love. He considers himself amply repaid if he feels that he added, be it ever so little, to the luster of the brightest jewel in the intellectual crown of mankind—The Science of Mathematics.

E · Mathematics—the Patrimony of the Race   Our discussion has thus led us to ascribe less importance to the role of the individual in the development of mathematics and to give more credit for the creation of this magnificent edifice to the human race as a whole. To be sure, it is always through the gifted individuals that the progress takes place. But no individual is indispensable in this task of furthering mathematical knowledge. The human race produces enough ability of a high degree to make the progress independent of any individual. Albert Einstein said in a press interview: "Individual worship, as I look at it, is always something unjustified. To be sure, nature does distribute her gifts in rich variety among her children. But of those richly gifted ones there are, thank God, many, and I am firmly convinced that most of them lead a quiet unobtrusive existence."

Mathematics is the patrimony of the human race. It is

the result of slow and patient labor of countless generations over a period of a great many centuries. Various practical callings have contributed towards this accumulation of mathematical knowledge and have furthered its development in the early and difficult stages. Modern technology provides such stimulation at an ever accelerating pace. The effort which has been expended in erecting the stately and imposing structure which we call mathematics is enormous. But mathematics has repaid the race for this effort. The practical value of mathematics cannot be overemphasized. To those privileged to appreciate the intellectual greatness of mathematics, the contemplation of this grandeur is an endless source of pure joy. The esthetic appeal of mathematics has found its enthusiastic and eloquent exponents. It would be proper to mention here another phase of the merit and value of mathematics to mankind.

The superiority of the human race over all the creatures inhabiting the earth, the reason that mankind is the master of this globe is due primarily to the fact that the experience of each generation does not die with that generation, but is transmitted to the next. This transmission of accumulated experience from generation to generation is the real power of the race, its greatest asset, its most powerful weapon in the conquest of nature, its surest tool in the accumulation of intellectual treasures. Nowhere is this more manifest than in mathematics. The cumulative character of mathematics is really astonishing. There is little in mathematics that ever becomes invalid, and nothing ever gets old. We may have all sorts of non-Euclidean-geometries, non-Archimedian-geometries, n-dimensional geometries, but all this makes the venerable elements of Euclid neither invalid nor obsolete. They remain, graceful and solid, an object of studies as much as ever, all in their own right. This cumulative process, this constant enlargement and perfectability of mathematics is the most precious of its characters, for it has given to mankind the idea of progress, with a clearness and distinctness that nothing else can equal, let alone surpass.

C. J. Keyser in his book *Humanism and Science*[20] goes a step farther and points out that the idea of progress suggested by science, and particularly by mathematics, has reflected upon the race itself. It has given mankind the idea that human nature in its turn may be perfected, that with the growth of knowledge and improved living conditions the human race will keep on rising to greater and greater heights on the road toward civilization. Mathematics has given the human race not only the technical tools to bend nature to its uses, not only a great and unequalled storehouse of intellectual beauty and enjoyment, but it also has given mankind a faith in itself and its destinies, hope and courage to carry on this unceasing struggle for a better, more noble, and more beautiful life.

FOOTNOTES

[1] Cf. Chapter VII, Section 1b.

[2] *Handbook of South American Indians*, James H. Steward, editor, Vol. 5 (Smithsonian Institution, Washington, D. C., 1929) p. 614.

[3] For instance, *What Happened in History*, V. Gordon Childe, (Pelican Books, A 108, London and Baltimore).

[4] Cf. Chapter I, Section 1b.

[5] Ibid.

[6] See Chapter III, Section 3c.

[7] Cf. Chapter III, Section 3h; Chapter VI, Section 2e.

[8] See Chapter V, Section 2b.

[9] "Law in History", *American Historical Review*, Edward P. Cheyney, Vol. 29 (1923-1924), pp. 231-248.

[10] Cf. Chapter V, Section 1c.

[11] *History of Analytic Geometry*, Carl B. Boyer (Scripta Mathematica Studies, New York, 1956).

[12] *Outline of the History of Mathematics*, R. C. Archibald, (Mathematical Association of America, 1949), p. 99, note 280.

[13] "Are Inventions Inevitable?" William F. Ogburn and Dorothy Thomas, *Political Science Quarterly*, Vol. 37 (1922), p. 83.

[14] *History of Analytic Geometry*, op. cit.

[15] "The Superorganic", A. L. Kroeber, *The American Anthropologist*, Vol. 19 (1917), p. 201.

[16] "The Locus of Mathematical Reality. An Anthropological Footnote", Leslie A. White, *Philosophy of Science*, Vol. 14, No. 4 (October, 1947), p. 298.

[17] *Number—The Language of Science*, Tobias Dantzig (First edition, New York, 1930), pp. 195-196.

[18] Cf. Chapter II, Section 1c.

[19] *The Instinct of Workmanship*, Thorstein Veblen (New York, 1914).

[20] *Humanism and Science*, C. J. Keyser (Columbia University Press, New York, 1931).

Chapter III

THE LURE OF THE INFINITE

1 · THE VAGARIES OF THE INFINITE

A · No Largest Number Have you ever had the opportunity of watching a bright youngster mastering the mechanism of naming numbers? It is a worth while experience, both entertaining and instructive. After the child has learned to name the numbers, say, up to twenty, he readily notices that counting beyond that, namely, twenty-one, twenty-two, . . . twenty-nine, consists in repeating the names of the first nine numbers he knows so well already, with the word twenty preceding them.

When you supply him, at the proper moment, with the word thirty, he will continue the scheme to forty, and so on, until in great triumph he comes to one hundred. Now repeating all the names he knows already, in the same order, preceded by the word one hundred he arrives at the number two hundred, then three hundred, . . . a thousand. One would naturally sympathize with the youngster in his feeling of achievement.

But he is not likely to rest on his laurels for very long. The child will with little or no help push forward catching on more and more readily to the nature of the almost automatic mechanism of advancing on the road towards the names of larger and larger numbers.

Of course, this will not happen all in one day, or one month. It may take a year, or more. In the meantime our youngster may learn, without much effort, the symbols we use to represent numbers, and names of numbers like millions, billions, trillions . . . . This mechanical way of extending the range of the names and symbols of numbers will finally lead a bright youngster to raise the inevitable question: where does it stop? where is the end of it? And at a tender age he will thus come to the realization that there is no largest number, that the series of integers is endless. It has a beginning, but no end. The little fellow has no difficulty in understanding that. The notion reduces itself to the simple idea that whatever the number, you can add one more unit and you have a larger number. A number is like a bus: nobody ever doubts that "there is always room for one more."

B · A Part as Big as the Whole   It is nearly beyond belief that a notion that seems to be within the grasp of a child should have baffled the greatest minds among both mathematicians and philosophers all through the ages. But this is literally the case. For this quite innocent looking "endless" series of numbers conceals behind its simple appearance many a joker that cannot readily be disposed of. Let us consider some of them.

Suppose we write down in a row the natural series of numbers, and directly underneath each number we put down its double, I mean the same number multiplied by two, like this:

1 2 3 4 5 6 7 8 9 10 11 12 13 14 . . .
2 4 6 8 10 12 14 16 18 20 22 24 26 28 . . . .

No matter how many numbers we may have in the first row, we will have just as many in the second. But this is absurd, for we know perfectly well that the even numbers also appear in the first row as a part of the series of natural numbers. Then where is the catch? The example given is one of the mildest possible. We could make matters worse by writing in the second row the numbers of the first row multiplied by three, or by seven, or by thirty-seven, which would make the absurdity more pronounced. A still more striking example is obtained if we write in the second line the squares of the numbers of the first line, so that the second line will consist of the numbers

$$1, 4, 9, 16, 25, 36 \ 49 \ldots.$$

We would thus be led to the conclusion that there are just as many perfect squares as there are numbers in the natural series of numbers. The "absurdity" of the conclusion is the more embarrassing in that between two consecutive perfect squares n^2 and $(n+1)^2$ there are $2n$ numbers which are not perfect squares. Thus between the square of 500,000 and the square of 500,001 there are a million numbers which are not perfect squares. If instead of squares we take cubes, fourth powers, and so on, matters are going from bad to worse all the time.

Similar difficulties, when the infinite is involved, are met with in geometrical considerations. Let us draw a fairly large segment AB (Fig. 2) of a straight line and another segment CD, considerably shorter. Now join the points A and C, B and D, and let the lines AC, BD meet in the point M. If we take any point, say E, on the segment AB and join it to M, the line ME will meet the segment CD in a point, say, F. If we reverse the order of operations and start with a point, J, on the segment CD, the line MJ will mark off a point I on AB. We can thus match every point of CD with a point of AB. It is natural to conclude from these constructions that there are as many points on the segment AB as there are on the segment CD. But this is a puzzling, indeed an absurd conclusion, for we have deliberately taken the segment AB considerably longer than CD. How is it that the number of points they contain are equal?

We could go on piling up such difficulties, accumulating such embarrassing conclusions. But this may become somewhat monotonous, and perhaps a little uncomfortable. So instead of adding new troubles it may be better to try to get out of

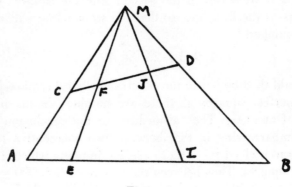

Figure 2

the troubles we are in already. Is there a way out of the difficulties we encountered? We may perhaps be able to cope with the problem by taking heed of the way we got into that mess. It is all due to the youngster whom we watched a while back, when he so complacently accepted the idea that the series of natural numbers is endless. That was reckless on his part, reckless indeed. For collections of objects we have direct experience with are all finite. We have ten fingers on our hands, we have about one hundred thousand hairs on our head, we have one hundred and sixty-five million co-citizens in our country, and we have nearly a three hundred billion dollar national debt largely due to military expenditures. But whether small or large, these collections are finite.[1] What we know about collections of objects we learned on finite collections. All the reasoning to which we are accustomed is applicable only to finite collections, and all of it goes topsy-turvy when we try to apply it to infinite collections.

We may perhaps be able to see how this happens if we re-examine one of the examples we considered before. Let us take the row of integers in the first line and the row of the same integers multiplied by two in the second line. If we stop the first line at, say, 18, the lower row stops at 36. The upper row contains the first half of the lower row, but not the second half, from 20 on. The same will be true no matter where we stop the first row. Thus it is not true that the lower row is a part of the upper one. But stopping anywhere makes our collection a finite one. And we are not supposed to stop. But if we are not to stop, if our collection of numbers is to go on without end, it does not matter which number of the lower row we may take into consideration, sooner or later it will appear also in the upper row. It does not matter in the least that the upper row is always behind the lower one. Ultimately the upper row will catch up with any assigned place of the lower row. This is due to the fact that the process has no stopping place, that it has no ultimate end. We are thus led to say that the number of even integers is as large as the number of all integers and in the same time maintain that the even integers are only a part of all the integers.

We agreed a while ago that we have ten fingers on our two hands. Nobody in his senses will argue that we have as many fingers on one hand as we have on both hands. That is patently absurd. But the collection of fingers we are talking about is finite. The number of even integers seems to be half the total number of integers, but they are as numerous as all the integers put together. This is not absurd, for an infinite collection.

C · Arithmetical Operations Performed on the Infinite
Now that we are a little better acquainted with the nature of infinite collections of objects and with the difference between

finite collections and infinite collections, we may try to take a closer look at the behavior of infinite collections. To simplify the language, let us agree to replace the phrase "infinite collection of objects" by the single word "infinity." When we say "infinity" we will still mean "an infinite collection of objects," but we will use the simpler, the shorter designation.

Infinity plus infinity is obviously infinity. So is infinity plus a finite collection. On the other hand the sum of two finite collections is necessarily a finite collection. But how about the difference between two infinite collections? How much is infinity minus infinity? You may say that it is infinity, and you may be right. If in the infinite collection of integers we propose to cross out all the even integers, this amounts to subtracting an infinite number of members of the total collection. The remainder of the collection will consist of all the odd integers which are still infinite in number, which still form an inexhaustible collection. A given segment of a straight line is an infinite collection of points, and so is a part of that segment. If from the whole segment we take away a part, what remains is an infinite collection of points.

The above examples show that if we say that infinity minus infinity is equal to infinity, this statement may be true, but is it always true? The answer to this question is: No. Infinity minus infinity may be a finite number, say, seven. A simple example may perhaps convince you of that. Suppose that in the natural series of integers we propose to strike out all the integers larger than seven. We would thus subtract an infinite collection from an infinite collection. And what would be the difference? Seven, exactly seven. If you would rather have the difference equal to another number, say 5 or 13, you are at perfect liberty to have it your way.

If a sum involves an infinite number of terms, each of them of finite size, is the sum finite or infinite? If you were offered the choice on a bet, which side would you take? There is a saying, attributed to Napoleon, that if two men engage in a bet, one of them is a crook, and the other man a fool. Should one offer you a bet on the answer to that question, he would definitely be the crook in the deal, for he would win no matter which side you chose.

That there are cases in which the sum of an infinite number of finite terms is infinite is clear enough. Take the number three and keep on adding it to itself. You will obtain the sums 3, 6, 9, 12, . . . and this row of numbers has no largest term, that is to say that the sum sought is infinite. But there are cases in which the addition of an infinite number of terms gives a result that is finite. A little story may bring out the point.

John, age ten, bought a pound of cherries from the neighborhood grocer, placed himself comfortably on the stairs of the back porch and had a feast. When he was all through and on the way to the garbage can with the collected pits he suddenly had a brilliant idea. He ran back to the grocer and told the man that he felt cheated. He bought cherries, but he has no use whatsoever for pits. The grocer, for reasons of his own, gave John half a pound of cherries in exchange for his pits. The boy disposed of the cherries in the same way as before, and he came back to the grocer with the pits, for which he got a quarter of a pound of cherries. If John keeps this game up indefinitely, how many pounds of cherries will he get from the grocer, cherries, mind you, not pits? In other words, let us now consider what is the ultimate value of the sum

$$\tfrac{1}{2}+\tfrac{1}{4}+\tfrac{1}{8}+\tfrac{1}{16}+\tfrac{1}{32}+ \ldots$$

if we should keep on adding terms indefinitely according to the same rule or scheme?

The answer may not seem quite clear, but it becomes obvious if we put the question in a different form. Let AB be a segment of a straight line one foot long, and let C be its mid-point (Fig. 3). To the segment AC add half the segment CB, thus forming the segment AD. To AD add half of DB to form the segment AE, and so on. This geometric procedure performs the addition of John's cherry problem. We may leave open the question whether we will reach the point B by this method. But what is quite clear is that on the one hand we get as close to B as we wish, if we keep the process up long enough, and on the other hand we will never get beyond B. Poor John, his cherry racket will never yield him more than one miserable pound of cherry meat, if that much.

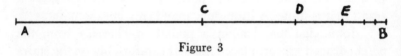

Figure 3

One may be inclined to argue that these problems are purely artificial, that mathematicians just invent them in order to have something to worry about, or something to talk about to each other, or perhaps to write books about. But actually it is fairer to say that problems of this sort are thrust upon them and the poor mathematicians do the best they can with a tough assignment. Surely, there is nothing extraordinary about a fraction like $\frac{1}{3}$, and it is nothing out of the way to attempt to convert it into a decimal fraction. The result is 0.333... where the number of decimals continues indefinitely. Now this decimal fraction may be written in the form

$$3/10 + 3/100 + 3/1000 + 3/10000 + \ldots.$$

and we are thus led to consider a sum with an endless number of terms. Obviously the ultimate value, or the limit of this sum, as mathematicians like to call it among themselves, has to be one third, the innocent little fraction we started with.

D · No Escape from the Infinite Such are some of the vagaries of the infinite. You may perhaps feel somewhat disturbed, or just a bit puzzled. But you need not let that worry you: you are in very good company. Since ancient times the most profound thinkers have struggled with those questions. In ancient Greece the problems raised by the consideration of the infinite were focused in the famous arguments of Zeno. Light-of-foot Archilles, Zeno argues, can never catch up with the proverbially slow moving turtle; or an arrow, the fastest moving thing known to antiquity, cannot move at all and must always remain in the same spot.[2]

A simple way of getting out of the trouble is to avoid the consideration of the infinite. Just give it up as a bad job. But this is much more readily said than done. We have already considered the problem of representing $\frac{1}{3}$ as a decimal fraction as an example of the way the infinite has of imposing itself. All of the calculus, that powerful tool of the mathematician, the physicist, the engineer, is squarely based on considerations of the infinite. The French mathematician Henri Poincaré (1854-1912), one of the greatest minds of all ages, said explicitly: "There can be no science but of the infinite."

2 · THE INFINITE IN GEOMETRY

A · Parallelism in Euclid's Elements   Parallel lines and parallel planes are all around us practically all the time. The opposite walls of the rooms where we spend such a large part of our days and nights are parallel, and the four corner lines of those rooms are parallel to each other. Two opposite edges of the tops of our tables are usually parallel lines, and so are the legs which support those tops. Many of the streets we walk on in our towns are parallel, and the two curbings on the two opposite sides of a street between which we drive our cars, are parallel lines. We plant many of our crops in parallel rows, and so on and on. There is hardly a notion more familiar to us than parallelism. Nevertheless, this seemingly innocent and harmless thing has been for the professional mathemati-

cian the *"enfant terrible"* of geometry, since the time of
Euclid, and most likely even before that.

Euclid based his theory of parallelism on a definition,
the last of the thirty five which he lists at the opening of his
Elements, and on a theorem (prop. 29, Book I), one of the
clumsiest in the book. Those two propositions were the first
to which objections were raised. The attack on them came
very early. It continued through the ages in various forms
until it was discovered that this particular theorem of Euclid's
was dissimulating behind its unattractive exterior nothing less
than non-Euclidean Geometry, and that the definition referred
to was trying to dodge the question of the infinite in geometry.

Here is Euclid's definition: "Parallel straight lines are
such as are in the same plane, and which, being produced
ever so far both ways, do not meet." This is quite simple, to
be sure, and plausible enough. But the statement seems to
invite a very pertinent, if obvious, question: How far is "ever
so far?" ten feet? a hundred yards? a thousand miles? a
million light-years? Euclid himself is careful not to raise this
question. Perhaps because he was aware of the skeleton in his
closet—the infinite. But a time came when the door of that
closet was thrown wide open. That valiant deed was accomplished by Projective Geometry, during the first quarter of
the 19th century, to say nothing about its precursors.

B · The Difference Between Metric and Projective Geometry
Projective Geometry[3] starts out with the same basic materials
as does Euclidean geometry, namely, points, lines, planes,
triangles, etc. But the two geometries emphasize two different
kinds of properties of the figures considered. For instance, if
Euclid comes across a couple of triangles, he inquires whether
a side of one of them happens to be as long as one of the sides
of the other triangle, and if so, whether the angles adjacent
to those two equal sides in the two triangles are respectively
equal. If this, too, happens to be the case, Euclid draws the

conclusion that the remaining sides of the two triangles are respectively equal.

Projective Geometry also takes an interest in the two triangles, but in a different way. In Projective Geometry we would join a vertex, say, A (Fig. 4) of the first triangle ABC to a vertex, say, A' of the second triangle A'B'C', a second vertex B of the first to a second vertex, say, B' of the second, and finally draw the line CC'. Now if it should happen that the

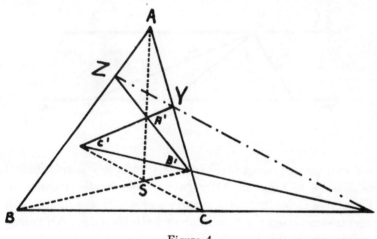

Figure 4

three lines AA', BB', CC' meet in the same point, say S, this fact would justify the following conclusion: if X denotes the point of intersection of the two sides BC, B'C' of the two triangles opposite the vertices A, A', and if Y, Z are similarly the points of intersection of the pairs of sides CA, C'A' and AB, A'B', respectively, the three points X, Y, Z are alined, that is, they lie on the same straight line. The point S and the line XYZ are said to be the center and the axis of perspectivity of the two triangles ABC, A'B'C'.

It is clear from this example that Euclid is primarily interested in the size of the elements of his figure. Projective

Geometry, on the other hand, ignores the metrical aspect of the figure and concentrates its attention on the relative position of the elements of the geometrical figure.

The difference of approach and interest of the two geometries accounts for the fact that Euclid got along without considering the infinite, while Projective Geometry had to face the music squarely. More than that, as we shall see presently, Euclid could not bring in the infinite without harming the coherence of his monumental work. It may also be pointed

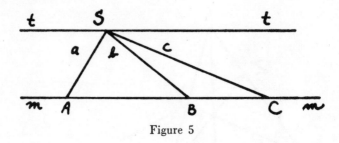

Figure 5

out that the nature of this beast—the infinite—is such that it is difficult to tame it. After having been in the harness of Projective Geometry for a century and a half, the infinite still leaves some room for discussion and clarification, even among authorities in this branch of mathematics.

C · The Point at Infinity of a Line   Projective Geometry in the plane considers two fundamental forms: the range of points (A, B, C, . . .) situated on a straight line m, and the pencil of rays a, b, c, . . . passing through the same point S (Fig. 5). Given the range m, we obtain a pencil S, if we join the points of m to a point S (not on m). Thus through every point of m there will pass a ray of S; conversely, every ray of S passes through a point of m, every ray, that is, but one, namely the ray t passing through S and parallel to m. This is a troublesome exception, worthy of further scrutiny.

The two lines m and t do not have a point in common.

Does that mean that they have nothing in common? A line, in addition to the many points that it has, possesses also an additional quality or property which we call "direction." The two lines m and t have this quality in common: they have the same direction. We could therefore make the statement that a ray of the pencil S has either a point or the direction in common with the line m.

The famous postulate of Euclid concerning parallel lines may be stated as follows: Through a given point one and only one line can be drawn having a giving direction. Hence the usual statement that "a line is determined by two of its points" may be supplemented to read: "or by one point and the direction of the line." Thus in the determination of a line the direction of the line plays the role of a point.

These remarks make it clear that the difficulty we encountered in connection with the range of points and the pencil of lines can readily be removed by identifying "direction" with a point. We can eliminate from our geometrical language the word "direction" and endow the line, in addition to all the "ordinary" points that it has, with a new "extraordinary" point. We will thus be able to make the statement that a line through S meets m in a point. In certain cases we may have to inquire whether the common point is an ordinary or an "extraordinary" point, i.e., whether we are dealing with a case of intersecting lines or of parallel lines. But in general, we will pay no attention to this distinction, not any more than we pay, in algebra, to the question as to whether a is greater than b when we write $a - b$. Our "extraordinary" point is usually called the "point at infinity" of the line. This name is justified on the ground that the point of intersection of a line through S with the line m keeps on receding indefinitely from any fixed point on m (say the foot of the perpendicular from S to m) as the line through S approaches the limiting position of parallelism with the line m. Some authors refer to this point as the "improper" point of the line, while others go to the opposite extreme and call it the "ideal" point of the line.

D · The Line at Infinity of a Plane and the Plane at Infinity of Space   In space, given a plane μ (mu) and a point S (Fig. 6), any line v of μ and the point S determine a plane; conversely, every plane passing through S cuts the plane μ along a line, every plane, that is, except one, namely the plane λ (lambda) through S which is parallel to μ. Here again it is not correct to say that since the planes μ and λ have no line in common, they have nothing in common. The two planes have the same "direction," or let us better say the same "orientation," to avoid overworking the same term and to take advantage of the abundance of words in the English language.

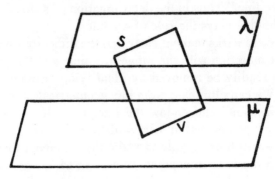

Figure 6

Through a point one and only one plane can be drawn parallel to a given plane. This proposition may be restated by saying: "A plane is determined by a point and the orientation of the plane." On the other hand, a plane is determined by a point and a line. Hence in the determination of a plane, the orientation of the plane plays the same role as a line. We can thus eliminate the exception noted, if we drop from our geometric vocabulary the word "orientation" and in its place endow the plane with an "extraordinary" line which we may call the "line at infinity," or the "improper" line, or the

"ideal" line of the plane. This convention enables us to say
that a plane through S always cuts the plane μ along a straight
line. Occasionally we may again have to inquire as to whether
this line is a line in the ordinary sense, or the fictitious line,
i. e., whether we are considering intersecting planes or parallel
planes. But in general we have no concern about this distinction. If it were otherwise, the whole scheme would serve
no useful purpose.

To return to plane geometry, the introduction of the point
at infinity freed us from a certain inconvenience. But this new
point raises troublesome questions of its own. Do the points

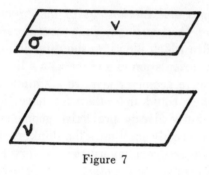

Figure 7

at infinity form a locus, that is, some kind of a figure, and if
so, what is that figure? The difficulty, however, is more apparent than real. Since every line in the plane has one and only
one point at infinity, the locus of these points, if there be
such, must be met by every line in the plane in one and only
one point; hence that locus can only be a straight line, the
"line at infinity" of the plane. This is a very fortunate circumstance, since it happens to agree with the "line at infinity" we
attributed to the plane when considering the plane in space.
This concordance is further strengthened by the consideration
of a line and a plane ν (Nu) parallel to each other.
Through (Fig. 7) a plane σ (sigma) may be drawn parallel

to ν, and the point of intersection of σ and lies on the line common to σ and ν, i.e., the line meets ν on the line at infinity of ν. Indeed, the point at infinity of the line belongs to the line at infinity of the plane ν, and the latter line is common to the two planes σ and ν.

We may ask, by analogy with the case of the plane, what is the locus in space containing all the lines at infinity of all the planes in space? The answer is based on the consideration that the locus must be a geometric entity with which every plane in space has a line in common and only one; hence that entity must itself be a plane, the "plane at infinity" of space.

E · *Advantages and Limitations of the Elements at Infinity* Projective Geometry enjoys a considerable advantage from the artifice which identifies the direction of a line with a point and the orientation of a plane with a line. The propositions of projective geometry acquire a simplicity and a generality that they could not otherwise have. Moreover, the elements at infinity give to projective geometry a degree of unification that greatly facilitates the thinking in this domain and offers a suggestive imagery that is very helpful in the acquisition of results. On the other hand, projective geometry stands ready to abandon these fictions whenever that seems desirable, and to express the corresponding propositions in terms of direction of a line and the orientation of a plane, to the great benefit of the science of geometry.

But the suggestive power of words is such that we are tempted to forget the precise and severe limitations under which the elements at infinity have been introduced. We are prone to ascribe to the elements at infinity other properties of points, lines, and planes. To take but one example: one might speculate on the implications of the fact that space is limited by a plane. To declare space limitless and to provide that limitless space with a boundary is sheer contradiction, at least in terms. On top of that, to claim that the statement is justified mathematically is utterly unfair. The plane at infinity of projective geometry has no mystical properties; it is simply a figure of speech, a round-about way of saying that through a given point one and only one plane can be drawn parallel to a given plane. Competent mathematicians do not take the "elements at infinity" of projective geometry for anything more than the convenient fiction that they are, within the limits of applicability of these elements, and do not hesitate to forsake them for something else that may prove to be more convenient under different circumstances.

F · Could Euclid Find Room in His Elements for Points at Infinity?   A much more serious problem arises when it is attempted to introduce the elements at infinity of projective geometry into the metrical geometry of Euclid. The proximity of the two branches of geometry acts as a powerful temptation. If a line has a point at infinity in projective geometry, why not in Euclidean geometry? Indeed, why did not Euclid himself think of the trick? A little reflection will show that the elements at infinity would work havoc with metrical geometry. At every point of a line a perpendicular can be erected and only one. Can a perpendicular be erected at the point at infinity of the line? Such a perpendicular does not exist, or at best is indeterminate, and the Euclidean proposition considered loses its generality. Two perpendiculars to the same line are parallel. If two parallel lines have a point at infinity in common, this contradicts the fundamental proposition that from a point outside a given line one and only one perpendicular can be drawn to the line. The points at infinity would ruin the entire theory of congruence of triangles. What would be the distance between two points on the line at infinity? If the answer is to be infinity, then every point of the line would be equidistant from all the other points on the line. And so on.

G · Do the Elements at Infinity "Enrich" Projective Geometry? Some writers on projective geometry insist that the straight line of projective geometry is the Euclidean straight line with an extra point added. The projective plane and projective space are, in turn, the Euclidean plane and Euclidean space enriched, respectively, by an additional line and an additional plane, just as nowadays, let us say, bread is enriched by added vitamins. These statements, despite their widespread acceptance, are nevertheless misleading. The extra point which projective geometry claims to add to the Euclidean line is the way in which projective geometry accounts for the property of the straight line which Euclidean geometry recognizes as the "direction" of the line. The difference between the Euclidean line and the projective line is purely verbal. The geometric content is the same.

Equally illusory is the difference between the Euclidean plane and the projective plane. The line at infinity of the projective plane is the way in which projective geometry incorporates into its plane geometry the parallelism of the Euclidean plane. The difference in verbiage may be striking, but the geometric substance is the same, and there is no justification for the claim of projective geometry that its plane is "richer" than the Euclidean plane. The same considerations obtain for the plane infinity. Euclidean space has its parallel planes. It suits the convenience of projective geometry to change the terminology and refer to this parallelism of Euclid by speaking of a plane at infinity; but such a change in nomenclature does not constitute an increase in geometric content.

The claim of having "enriched" Euclidean space has not led projective geometry to make any unjustifiable use of its truly marvelous elements at infinity. It is nevertheless desirable that we dot our i's and know precisely the origin and relation of these elements in the two geometries. It makes for clearer thinking. It may also help to dispel some of the fog of mathematical mysticism.

3 · THE MOTIONLESS ARROW

A · Arrows In the lore of mankind the arrow occupies a
conspicuous place, a place of distinction. There is the heroic
arrow with which the legendary William Tell, at the behest
of a tyrant, shot an apple off his own son's head, to say
nothing of the other arrow that Tell held in reserve for the
tyrant himself, in case his first aim should prove too low.
There is the soaring first arrow of Hiawatha that would not
touch the ground before the tenth was up in the air. There
is the universally famous romantic arrow with which Cupid
pierces the hearts of his favorites—or shall I say victims?

There is also an arrow that is philosophical, or scientific,
or, better still, both. This famous "motionless arrow," as it
may best be called, has stirred the mind, excited the imagination, and sharpened the wits of profound thinkers and erudite
scholars for well over two thousand years.

B · The Arguments of Zeno and Those of His Imitators
Zeno of Elea, who flourished in the fifth century B.C., confronted his fellow philosophers and anybody else who was
willing to listen with the bold assertion that an arrow, the
swiftest object known to his contemporaries, cannot move at
all.

According to Aristotle, Zeno's argument for, or proof of,
his embarrassing proposition ran as follows: "Everything,
when in uniform state, is continually either at rest or in motion, and a body moving in space is continually in the Now
(instant), hence the arrow in flight is at rest." Some six centuries later another Greek philosopher offered a somewhat
clearer formulation of the argument: "That which moves
can neither move in the place where it is, nor yet in the
place where it is not." Therefore, motion is impossible.

The "motionless arrow" was not Zeno's only argument

of its kind. He had others. Zeno had Achilles engage in a race with a tortoise and showed a priori that the "light-of-foot" Achilles could never overtake the proverbially slow turtle. In Aristotle's presentation, here is the argument: "In a race the faster cannot overtake the slower, for the pursuer must always first arrive at the point from which the one pursued has just departed, so that the slower is always a small distance ahead." A modern philosopher states the argument more explicitly: "Achilles must first reach the place from which the tortoise has started. By that time the tortoise will have got on a little way. Achilles must then traverse that, and still the toroise will be ahead. He is always nearer, but he never makes up to it."

A third argument of Zeno's against motion is known as the "Dichotomy." In Aristotle's words: "A thing moving in space must arrive at the mid-point before it reaches the end-point." J. Burnet offers a more elaborate presentation of this argument:

> You cannot traverse an infinite number of points in a finite time. You must traverse half a given distance before you traverse the whole, and half of that again before you traverse it. This goes on ad infinitum, so that (if space is made up of points) there are an infinite number in any given space, and it cannot be traversed in a finite time.

Zeno had still other arguments of this kind. But I shall refrain from quoting them, for by now a goodly number of you have no doubt already begun to wonder what this is all about, what it is supposed to mean, if anything, and how seriously it is to be taken. Your incredulity, your skepticism, reflect the intellectual climate in which you were brought up and in which you continue to live. But that climate has not always been the same. It has changed more than once since the days of Zeno.

To take a simple example. We teach our children in our schools that the earth is round, that it rotates about its axis, and also that it revolves around the sun. These ideas are an integral part of our intellectual equipment, and it seems to us impossible to get along without them, much less to doubt them. And yet when Copernicus, or Mikolaj Kopernik, as the Poles call him, published his epoch-making work barely four centuries ago, in 1543, the book was banned as sinful. Half a century later, in 1600, Giordano Bruno was burned at the stake in a public place in Rome for adhering to the Copernican theory and other heresies. Galileo, one of the founders of modern science, for professing the same theories, was in jail not much more than three centuries ago.

What Zeno himself thought of his arguments, for what reasons he advanced them, what purpose he wanted to achieve by them, cannot be told with any degree of certainty. The data concerning his life are scant and unreliable. None of his writings are extant. Like the title characters of some modern novels such as *Rebecca*, by Daphne du Maurier, or *Mr. Skeffington*, by Elizabeth Arnim Russell, Zeno is known only by what is told of him by others, chiefly his critics and detractors. The exact meaning of his arguments is not always certain.

Zeno may or may not have been misinterpreted. But he certainly has not been neglected. Some writers even paid him the highest possible compliment—they tried to imitate him. Thus the "Dichotomy" suggested to Giuseppe Biancani, of Bologna, in 1615 a "proof" that no two lines can have a common measure. For the common measure, before it could be applied to the whole line, must first be applied to half the line, and so on. Thus the measure cannot be applied to either line, which proves that two lines are always incommensurable.

A fellow Greek, Sextus Empiricus, of the third century A.D., taking the "motiónless arrow" for his model, argued that a man can never die, for if a man dies, it must be either at a time when he is alive or when he is dead, etc.

It may be of interest to mention in this connection that the Chinese philosopher Hui Tzu argued that a motherless colt never had a mother. When it had a mother it was not motherless and at every other moment of its life it had no mother.

C · Aristotle's Arguments About the Infinite Divisibility of Both Time and Space   Some writers offered very elaborate interpretations of Zeno's arguments. These writers saw in the creator of these arguments a man of profound philosophical insight and a logician of the first magnitude. Such was the attitude of Immanuel Kant and, a century later, of the French mathematician Jules Tannery (1848-1910). To Aristotle, who was born about a century after Zeno, these arguments were just annoying sophisms whose hidden fallacy it was all the more necessary to expose in view of the plausible logical form in which they are clothed. Other writers displayed just as much zeal in showing that Zeno's arguments are irrefutable.

Aristotle's fundamental assumptions are that both time and space are continuous, that is, "always divisible into divisible parts." He further adds: "The continual bisection of a quantity is unlimited, so that the unlimited exists potentially, but it is never reached."

With regard to the "arrow" he says:

A thing is at rest when it is unchanged in the Now and still in another Now, itself as well as its parts remaining in the same status. . . . There is no motion, nor rest in the Now. . . . In a time interval, on the contrary, it (a variable) cannot exist in the same state of rest, for otherwise it would follow that the thing in motion is at rest.

That it is impossible to traverse an unlimited number of half-distances (the "Dichotomy"), Aristotle refutes by pointing out that "time has unlimitedly many parts, in consequence of which there is no absurdity in the consideration that in an unlimited number of time intervals one passes over unlimited many spaces." The argument Aristotle directs against "Achilles" is as follows:

> If time is continuous, so is distance, for in half the time a thing passes over half the distance, and, in general, in the smaller time the smaller distance, for time and distance have the same divisions, and if one of the two is unlimited, so is the other. For that reason the argument of Zeno assumes an untruth, that one unlimited cannot travel over another unlimited along its own parts, or touch such an unlimited, in a finite time; for length as well as time and, in general, everything continuous, may be considered unlimited in a double sense, namely according to the (number of) divisions or according to the (distances between the) outermost ends.[4]

Aristotle seems to insist that as the distances between Achilles and the tortoise keep on diminishing, the intervals of time necessary to cover these distances also diminish, and in the same proportion.

The reasonings of Aristotle cut no ice whatever with the French philosopher Piere Bayle (1647-1706), who in 1696 published his *Dictionnaire Historique et Critique*, translated into English in 1710. Bayle goes into a detailed discussion of Zeno's arguments and is entirely on the side of Zeno. He categorically rejects the infinite divisibility of time.

> Successive duration of things is composed of moments, properly so called, each of which is simple and indivisible, perfectly distinct from the

past and future and contains no more than the
present time. Those who deny this consequence must
be given up to their stupidity, or their want of sin-
cerity, or to the unsurmountable power of their
prejudices.

Thus the "Arrow" will never budge.

The philosophical discussion of the divisibility or the
nondivisibility of time and space continues through the centuries. As late as the close of the past century Zeno's arguments based on this ground were the topic of a very animated
discussion in the philosophical journals of France.

D · The Potential Infinity and the Actual Infinity A
mathematical approach to "Achilles" is due to Gregory St.
Vincent (1584-1667), who in 1647 considered a segment

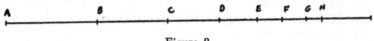

Figure 8

AK on which he constructed an unlimited number of points
B,C,D, . . . such that $AB/AK = BC/BK = CD/CK = . . . = r$,
(Fig. 8), where r is the ratio, say of the speed of the tortoise
to the speed of Achilles. He thus obtains the infinite geometric
progression $AB + BC + CD + . . .$ and, since this series is con-
vergent, Achilles does overtake the elusive tortoise.

Descartes solved the "Achilles" by the use of the geometric progression $1/10 + 1/1001/1000 + . . . = 1/9$. Later
writers quoted this device or rediscovered it time and again.
But this solution of the problem raised brand-new questions.

St. Vincent overlooked the important fact that Achilles
will fail to overtake the slow-moving tortoise after all, unless
the variable sum of the geometric progression actually reaches
its limit. Now: Does a variable reach its limit, or does it
not? The question transcends, by far, the "Achilles." It was,
for instance, hotly debated in connection with the then nascent

differential and integral calculus. Newton believed that his variables reached their limits. Diderot (1713-1784), writing a century or so later in the famous *Encyclopédie*, is quite definite that a variable cannot do that, and so is A. De Morgan (1806-1871), in the *Penny Cyclopedia* in 1846. Sadi Carnot (1796-1832) and A. L. Cauchy (1789-1857), like Newton, have no objection to variables reaching their limits.

The other question that arises in connection with St. Vincent's progression is: How many terms does the progression have? The answer ordinarily given is that the number is infinite. This answer, however, may have two different meanings. We may mean to say that we can compute as many terms of this progression as we want and, no matter how many we have computed, we can still continue the process. Thus the number of terms of the progression is "potentially" infinite. On the other hand, we may imagine that all the terms have been calculated and are all there forming an infinite collection. That would make an "actual" infinity. Are there actually infinite collections in nature? Obviously, collections as large "as the stars of the heaven, and as the sand which is upon the seashore," are nevertheless finite collections.[5]

From a quotation of Aristotle already given it would seem that he did not believe in the actually infinite. Galileo, on the other hand, accepted the existence of actual infinity, although he saw clearly the difficulties involved. If the number of integers is not only potentially but actually infinite, then there are as many perfect squares as there are integers, since for every integer there is a perfect square and every perfect square has a square root.[6] Galileo (1564-1642) tried to console himself by saying that the difficulties are due to the fact that our finite mind cannot cope with the infinite. But De Morgan sees no point to this argument, for, even admitting the "finitude" of our mind, "it is not necessary to have a blue mind to conceive of a pair of blue eyes."

A younger contemporary of Galileo, the prominent English philosopher Thomas Hobbes (1588-1679), could not accept Galileo's actual infinity, on theological grounds. "Who thinks that the number of even integers is equal to the number of all integers is taking away eternity from the Creator." However, the very same theological reasons led an illustrious younger contemporary of Hobbes, namely, G. W. Leibniz (1646-1717), to the firm belief that actual infinities exist in nature *pour mieux marquer les perfections de son auteur*.

The actual infinite was erected into a body of doctrine by Georg Cantor (1845-1918) in his theory of transfinite numbers. The outstanding American historian of mathematics, Florian Cajori (1859-1930), considers that this doctrine of Cantor's provided a final and definite answer to Zeno's paradoxes and thus relegates them to the status of "problems of the past."

Tobias Danzig in his *Number, the Language of Science* is not quite so happy about it, in view of the fact that the whole theory of Cantor's is of doubtful solidity.

E · Motion and Dynamics Whatever may have been the reasons that prompted Zeno to promulgate his paradoxes, he certainly must have been a man of courage if he dared to deny the existence of motion. We learn of motion and learn to appreciate it at a very, very early age; motion is firmly imbedded in our daily existence and becomes a basic element of our psychological make-up. It seems intolerable to us that we could be deprived of motion, even in a jest.

Nevertheless, the systematic study of motion is of fairly recent origin. The ancient world knew a good deal about Statics, as evidenced by the size and solidity of the structures that have survived to the present day. But they knew next to nothing about Dynamics, for the forms of motion with which they had any experience were of very narrow scope. Their machines were of the crudest and very limited in variety. Zeno's paradoxes of motion were for the Greek philosophers "purely academic" questions.

The astronomers were the first to make systematic observations of motion not due to muscular force and to make
deductions from their observations. Man studied motion in
the skies before he busied himself with such studies on earth.
How difficult it was for the ancients to dissociate motion
from muscular effort is illustrated by the fact that Helios
(the sun) was said by the Greeks to have a palace in the
east whence he was drawn daily across the sky in a fiery
chariot by four white horses to a palace in the west.

The famous experiments of Galileo with falling bodies
are the beginning of modern Dynamics. The great voyages
created a demand for reliable clocks, and the study of clock
mechanisms and their motion engaged the attention of such
outstanding scholars as Huygens. No small incentive for
the study of motion was provided by the needs of the developing artillery. The gunners had to know the trajectories
of their missiles. The theoretical studies of motion prompted
by these and other technical developments were in need of
a new mathematical tool to solve the newly arising problems,
and calculus came into being.[7]

The infinite, the infinitesimal, limits and other notions
that were involved, perhaps crudely, in the discussion of
Zeno's arguments were also involved in this new branch of
mathematics. These notions were as hazy as they were essential. Both Newton and Leibniz changed their views on
these points during their lifetimes because of their own critical acumen as well as the searching criticism of their contemporaries. But neither of them ever entertained the idea of
giving up their precious find, for the good and sufficient
reason that this new and marvelous tool gave them the solution
of some of the problems that had defied all the efforts of
mathematicians of preceding generations. The succeeding
century, the eighteenth, exploited to the utmost this new
instrument in its application to the study of motion, and
before the century was over it triumphantly presented to the
learned world two monumental works: the *Mécanique Analy-*

tique of J. L. Lagrange (1736-1813), and the *Mécanique Céleste* of P. S. Laplace (1749-1827).

The development of Dynamics did not stop there. It kept pace with the phenomenal development of the experimental sciences in the nineteenth century. These theoretical studies on the one hand served as a basis for the creation of a technology that surpassed the wildest dreams of past generations and on the other hand changed radically our attitude toward many of the problems of the past; they created a new intellectual atmosphere, a new "intellectual climate."

Zeno's arguments, or paradoxes, if you prefer, deal with two questions which in the discussions of these paradoxes are very closely connected, not to say mixed up: What is motion, and how can motion be accounted for in a rational, intellectual way? By separating the two parts of the problem we may be able to come much closer to finding a satisfactory answer to the question, in accord with the present-day intellectual outlook.

F · Motion—An "Undefined Term"   The critical study of the foundations of mathematics during the nineteenth century made it abundantly clear that no science and, more generally, no intellectual discipline can define all the terms it uses without creating a vicious circle. To define a term means to reduce it to some more familiar component parts. Such a procedure obviously has a limit beyond which it cannot go. Most of us know what the color "red" is. We can discuss this color with each other; we can wonder how much the red color contributes to the beauty of the sunset; we can make use of this common knowledge of the red color for a common purpose, such as directing traffic. But we cannot undertake to explain what the red color is to a person born color blind.

In the science of Dynamics motion is such a term, such an "undefined" term, to use the technical expression for it.[8]

Dynamics does not propose to explain what motion is to

anyone who does not know that already. Motion is one of its starting points, one of its undefined, or primitive, terms. This is its answer to the question: What is motion?

You have heard many stories about Diogenes (412-323 B.C.). He lived in a barrel. He threw away his drinking cup when he noticed a boy drinking out of the hollow of his hand. He told his visitor, Alexander the Great, that the only favor the mighty conqueror could possibly do for him was to step aside so as not to obstruct the sun for the philosopher. Well, there is also the story that when Diogenes was told of Zeno's arguments about the impossibility of motion, he arose from the place where he was sitting on the ground alongside his barrel, took a few steps, and returned to his place at the barrel without saying a single word. This was the celebrated Cynic philosopher's "eloquent" way of saying that motion *is*. And did he not also say at the same time that motion is an "undefined term"?

St. Augustine (354-450) used an even more convincing method to emphasize the same point. He wrote:

> When the discourse (on motion) was concluded, a boy came running from the house to call for dinner. I then remarked that this boy compels us not only to *define* motion, but to see it before our very eyes. So let us go and pass from one place to another, for that is, if I am not mistaken, nothing else than motion.

The revered theologian seems to have known, from personal experience, that nothing is as likely to set a man in motion as a well-garnished table.

G · Theory and Observation   Let us now turn to the second part involved in Zeno's paradoxes, namely, how to account for motion is a rational way. All science may be said to be an attempt to give a rational account of events in nature, of the ways natural phenomena run their courses. The scientific theories are a rational description of nature that enables us to foresee and foretell the course of natural events. This characteristic of scientific theories affords us an intellectual satisfaction, on the one hand, and, on the other hand, shows us how to control nature for our benefit, to serve our needs and comforts. *Prévoir pour pouvoir*, to quote Henri Poincaré. A scientific theory, that is, a rational description of a sector of nature, is acceptable and accepted only as long as its previsions agree with the facts of observation. There can be no bad theory. If a theory is bad or goes bad, it is modified or it is thrown out completely.

"Achilles" is an attempt at a rational account of a race, a theoretical interpretation of a physical phenomenon. The terrible thing is that Zeno's theory predicts one result, while everybody in his senses knows quite well that exactly the contrary actually takes place. Aristotle in his time and day felt called upon to use all his vast intellectual powers to refute the paradox. Our present intellectual climate imposes no such obligation upon us. If saying that in order to overtake the tortoise Achilles must first arrive at the point from which the tortoise started, etc., leads to the conclusion that he will never overtake the creeping animal, we simply infer that Zeno's theory of a race does not serve the purpose for which it was created. We declare the scheme to be unworkable and proceed to evolve another theory which will render a more satisfactory account of the outcome of the race.

That, of course, is assuming that the theory of Zeno was offered in good faith. If it was not, then it is an idle plaything, very amusing, perhaps, very ingenious, if you like, but not worthy of any serious consideration. There are more worthwhile ways of spending one's time than in shadow boxing. Our indifferent attitude towards Zeno's paradoxes is perhaps best manifested by the fact that the article "motion" in the *Britannica* does not mention Zeno, whereas Einstein is given considerable attention; the *Americana* dismisses "motion" with the curt reference "see Mechanics."

Consider an elastic ball which rebounds from the ground to ⅔ of the height from which it fell. When dropped from a height of 30 feet, how far will the ball have traveled by the time it stops? Any bright freshman will immediately raise the question whether that ball will ever stop. On the other hand, that same freshman knows full well that after a while the ball will quietly lie on the ground. Will we be very much worried by this contradiction? Not at all. We will simply draw the conclusion that the law of rebounding of the ball, as described, is faulty.

H · Instantaneous Velocity   The difficulties encountered in connection with the question of a variable reaching or not reaching a limit are of the same kind and nature. The mode of variation of a variable is either a description of a natural event or a creation of our imagination, without any physical connotation. In the latter case, the law of variation of the variable is prescribed by our fancy, and the variable is completely at our mercy. We can make it reach the limit or keep it from doing so, as we may see fit. In the former case it is the physical phenomenon that decides the question for us.

Two bicycle riders, 60 miles apart, start towards each other, at the rate of 10 miles per hour. At the moment when they start a fly takes off from the rim of the wheel of one rider and flies directly towards the second rider at the rate of 15 miles per hour. As soon as the fly reaches the second rider it turns around and flies toward the first, etc. What is the sum of the distances of the oscillations of the fly? In Zeno's presentation the number of these oscillations is infinite. But the flying time was exactly 3 hours, and the fly covered a distance of 45 miles. The variable sum actually reached its limit.

The sequence of numbers 1, $\frac{1}{2}$, $\frac{1}{4}$, $\frac{1}{8}$, $\frac{1}{16}$, . . . obviously has for its limit zero. Does the sequence reach its limit? Let us interpret this sequence, somewhat facetiously, in the following manner. A rabbit hiding in a hollow log noticed a dog standing at the end near him. The rabbit got scared and with one leap was at the other end; but there was another dog. The rabbit got twice as scared, and in half the time he was back at the first end; but there was the first dog, so the rabbit got twice as scared again, etc. If this sequence reaches its limit, the rabbit will end up by being at both ends of the log at the same time.

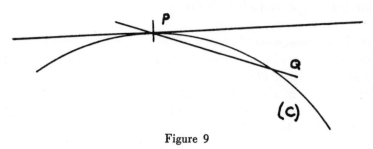

Figure 9

If a point Q of a curve (C) moves towards a fixed point P (Fig. 9) of the curve, the line PQ revolves about P. If Q approaches P as a limit, the line PQ obviously approaches as a limiting position the tangent to the curve (C) at the point P; and if the point Q reaches the position P or, what is the same thing, coincides with P, the line PQ will coincide with the tangent to (C) at P.[9]

If s represents the distance traveled by a moving point in the time t, does the ratio s/t approach a limit when t approaches zero as a limit? In other words, does a moving object have an instantaneous velocity at a point of its course, or its trajectory? Aristotle could not answer that question; he probably could not make any sense of the question. Aristotle agreed with Zeno that there can be no motion in the

Now (moment). But to us the answer to this question is not subject to any doubt whatever: we are too accustomed to read the instantaneous velocities on the speedometers of our cars.

The divisibility or the nondivisibility of time and space was a vital question to the Greek philosophers, and they had no criterion according to which they could settle the dispute. To us time and space are constructs that we use to account for physical phenomena, constructs of our own making, and as such we are free to use them in any manner we see fit. Albert Einstein did not hesitate to mix up the two and make of them a space-time continuum when he found that such a construct is better adapted to account for physical phenomena according to his theory of relativity.

I · A Modern Answer to Zeno's Paradoxes   We have dealt with the two parts of Zeno's paradoxes: the definition of motion and the description of motion. There is, however, a third element in these paradoxes, and it is this third element that is probably more responsible for the interest that these paradoxes held throughout the centuries than those we have considered already. This is the logical element.

That Zeno was defending an indefensible cause was clear to all those who tried to refute him. But how is it possible to defend a false cause with apparently sound logic? This is a very serious challenge. If sound logic is not an absolute guaranty that the propositions defended by that method are valid, all our intellectual endeavors are built on quicksand, our courts of justice are meaningless pantomime, etc.

Aristotle considered that the fundamental difficulty involved in Zeno's argument against motion was the meaning Zeno attached to his "Now." If the "Now," the moment, as we would say, does not represent any length of time but only the durationless boundary between two adjacent intervals of time, as a point without length is the common boundary of two adjacent segments of a line, then in such a moment there can be no motion; the arrow is motionless. Aristotle tried to refute Zeno's denial of motion by pointing out that it is wrong to say that time is made up of durationless moments. But Aristotle was not very convincing, judging by the vitality of Zeno's arguments.

Our modern knowledge of motion provided us with better ways of meeting Zeno's paradoxes. We can grant Zeno both the durationless "Now" and the immobility of the object in the "Now" and still contend that these two premises do not imply the immobility of the arrow. While the arrow does not move in the "Now," it conserves its capacity, its potentiality of motion. In our modern terminology, in the "Now" the arrow has an instantaneous velocity. This notion of instantaneous velocity is commonplace with us; we read it "with our own eyes" on our speedometers every day. But it was completely foreign to the ancients. Thus Zeno's reasoning was faulty because he did not know enough about the subject he was reasoning about.[10]

Zeno's apparently unextinguishable paradoxes, as they are referred to by E. T. Bell (b. 1883) in an article published in *Scripta Mathematica*, will not be put out of circulation by our remarks about them. I have no illusions about that; neither do I have any such ambitions. These paradoxes have amused and excited countless generations, and they should continue to do so. Why not?

FOOTNOTES

[1] Cf. Chapter III, Section 3d.
[2] Cf. Chapter III, Section 3.
[3] See Chapter I, Section 2d.
[4] Cf. Chapter III, Section 1c.
[5] See Chapter III, Section 1b.

[6] *Ibid.*
[7] Cf. Chapter II, Section 1c.
[8] Cf. Chapter I, Section 3f; Chapter V, Section 1j.
[9] Cf. Chapter II, Section 1c; Chapter V, Section 2c.
[10] Cf. Chapter I, Section 3e.

Chapter IV

MATHESIS THE BEAUTIFUL!

1 · MATHEMATICS AND ESTHETICS

A · Beauty in Mathematics A cultured person with literary proclivities once asked this writer whether mathematicians see beauty in their science. During her school career she has heard her teacher of mathematics, whose subject, by the way, she enjoyed very little, refer to a theorem as being beautiful, and this statement seemed to her preposterous. In reply to her question one could have quoted those great masters of mathematical thought who spoke eloquently on the subject. Henri Poincaré (1854-1912), one of the greatest mathematicians and one of the greatest minds of all times, said in this connection: "Above all, adepts find in mathematics delights analogous to those that painting and music give. They admire the delicate harmony of numbers and of forms; they are amazed when a new discovery discloses to them an unlooked for perspective, and the joy they thus experience has it not the esthetic character, although the senses take no part in it? Only the privileged few are called to enjoy it fully, but is it not so with all the noblest arts?" Our distinguished contemporary Bertrand Russell (b. 1872) said: "Mathematics,

rightly viewed, possesses not only truth, but supreme beauty
—a beauty cold and austere, like that of sculpture . . . The
true spirit of delight, the exaltation, the sense of being more
than a man which is the touchstone of the highest excellence,
is to be found in mathematics as surely as in poetry." Let me
add just one more quotation, from an American clergyman
and mathematician, Thomas Hill (1818-1893), "The mathematics is usually considered as being the very antipodes of
poesy. Yet mathesis and poesy are of the closest kindred, for
they are both works of imagination."

B · Mathematics in Beauty For the initiate mathematics
has very much in common with the fine arts. On the other
hand the fine arts are greatly indebted to mathematics. To
achieve verse rhythm the poet must *count* the feet in his lines,
i.e. the regularly recurring *metrical units*. The words in a
verse must be placed in *measured* and cadenced formation
so as to produce a *metrical* effect.

Arturo Aldunate Phillips, a Chilean essayist, economist,
poet, and engineer, in his *Matematica y Poesia* (*essayo y
entusiasmo*)[1] goes much farther. He not only sees many close
ties, both intellectual and artistic, between mathematics and
poetry, but he traces a close parallel between the two, in their
historical development, as well as in their role in the history
of culture.

The role of mathematics in music is a quite intimate one.
Several centuries before our present era Pythagoras observed
already that when musical strings of equal length are stretched
by weights having the proportions of $\frac{1}{2}$, $\frac{2}{3}$, $\frac{3}{4}$, they produce
intervals which are an octave, a fifth, and a fourth. Ever
since that time mathematicians have greatly contributed towards the elaboration of the theory of music. Euclid, the
author of the famous Elements, wrote two books on the theory
of music. When the music of the ancients, the homophonic
music, gave way to the polyphonic music of the middle ages,

mathematicians have furthered its theoretical development. The Renaissance has witnessed the birth of our modern, harmonic music, and among those who contributed towards the study of its theory we find such names as Kepler (1571-1630), Descartes (1596-1650), Huygens (1629-1695).

The close connection between mathematics and music has been expressed by H. Helmholtz (1821-1894) as follows: "Mathematics and music, the most sharply contrasted fields of scientific activity, and yet related, supporting each other, as if to show forth the secret connection which ties together all the activities of the mind, and which leads to surmise that the manifestations of the artist's genius are but unconscious expressions of a mysteriously acting rationality." Leibniz is even more specific: "Music is a hidden exercise in arithmetic, of a mind unconscious of dealing with numbers." The love of mathematicians for music is a well established fact. The great contemporary mathematical genius Albert Einstein (1878-1955) was an excellent violinist.

Sculpture, architecture, painting, and the graphic arts in general, obviously involve geometric considerations. What geometric constructions artists have used, consciously or unconsciously, to achieve their esthetic effects has been well analysed and put clearly into evidence. We shall mention that one of the most telling esthetic effects is obtained by the so-called "Golden Section" and its derivatives, and this section is connected with the quadratic equation $x^2 - ax = a^2$.[2] It is far from a mere coincidence that great artists like Leonardo da Vinci, Raphael, Michael Angelo and Albert Durer felt a very great attraction for mathematics. The great accumulation of knowledge of our own day makes such manifestations more rare.

Those not belonging to the fortunate few who can discern beauty in mathematics, may perhaps learn to perceive mathematics in beauty.

2 · ART AND MATHEMATICS

A · Mathematics, Logic, Music   Mathematics is proud, and justly proud, of the logic of its proofs. But these proofs are not an integral part of the mathematical doctrine. They are tools the mathematician uses in order to achieve his results, they are the vehicle he mounts to cover the territory he wants to survey, to contemplate, to admire. The car you travel in is not a part of the beautiful scenery your journey uncovers for you. True, to the people who have made only a slight and passing acquaintance with mathematics, this subject consists of nothing but proofs. They are in the same position as the traveler who has had to expend so much effort to make his vehicle go forward that he has no time and no eyes for the countryside. Another popular misconception is that mathematics consists of propositions, of so many chopped up sentences, of so many little pigeon-holes. To the initiated, however, mathematics is just as connected, just as continuous as a beautiful landscape. The propositions are singled out as points of orientation, as spots of particular interest or attraction, as elevations from which an unexpected, an unusually pleasing view may be beheld. But propositions are not the whole of mathematics any more than the elevations of the countryside constitute the whole view. A contemporary philosopher, Professor Scott Buchanan, said: "The structures with which mathematics deals are like lace, like the leaves of trees, like the play of light and shadow on a meadow, or on a human face."

Mathematical proofs may not be a part of the mathematical structure when the latter is completed, but they are one of the charms connected with the study of the subject. The best mathematical proofs are usually short, direct, and penetrating. The "indeed" of such a proof has sometimes the mellowness of a condescending smile, sometimes the swiftness of epigrammatic sarcasm and sometimes the surprise of a pointed, witty anecdote. A long mathematical proof may lack the directness of a short one, but often it makes up for it by having the swing and sway, the rhythm of music. It even may have the very structure of a musical composition. Here is a casual, seemingly irrelevant beginning of a train of thought carried to a certain point and dropped, like a musical theme developed to a certain degree and abandoned for another, apparently unrelated theme; another line of logical argument, and then perhaps also a third is started, in the same way. Then the separate logical arguments, the musical themes, begin to approach each other, to intermingle, to intertwine, then they become closely knit together, and finally burst out in a triumphant finale of achievement.

If you grant that there may be music in mathematical logic, then it is nothing but mathematical custom to inquire whether the converse of the proposition is not true, whether a musical composition may not resemble a logical reasoning. This may be too far fetched, but I, for one, am inclined to believe that this converse proposition is true. Take any masterpiece of musical literature, a short piece, for the sake of simplicity; say, Schubert's "Ave Maria," if you will. Forget the words of this song, think of it as played on the violin. It starts out with a brief phrase, a simple statement. The next phrase is more emphatic, but not yet sufficiently convincing. Follow a series of musical sentences, bound together by an indestructible tie of logical necessity, each more insistent, each overshadowing and enforcing the preceding one, until the highest pitch is reached—the irresistible argument of ecstasy, after which there returns the musical phrase of the beginning —the proposition is proved.

B · Opinions of Mathematicians About Mathematics   Whatever may be said for the beauty of mathematical logic, we may be expected to say something about the interrelation of mathematics proper and beauty. But, in the first place, is there any such relation? Is there not rather direct opposition? What, pray, can rigid, cold, calculating mathematics have in common with subtle, creative, lofty, imaginative art? This question faithfully mirorrs the state of mind of most people, even of most educated people. But the great leaders of mathematical thought, the creators and builders of the noble edifice of the oldest of sciences, have frequently and repeatedly asserted that the object of their pursuits is just as much an art as it is a science, and perhaps even outright a fine art.

The bearer of one of the greatest names among American mathematicians at the beginning of the century, Maxime Bôcher (1867-1918) wrote: "I like to look at mathematics almost more as an art than as a science; for the activity of the mathematician, constantly creating as he is, guided although not controlled by the external world of senses, bears a resemblance, not fanciful, I believe, but real, to the activities of the artist, of a painter, let us say. Rigorous deductive reasoning on the part of the mathematician may be likened here to the technical skill in drawing on the part of the painter. Just as one cannot become a painter without a certain amount of skill, so no one can become a mathematician without the power to reason accurately up to a certain point. Yet these qualities, fundamental though they are, do not make a painter or a mathematician worthy of the name, nor indeed are they the most important factors in the case. Other qualities of a far more subtle sort, chief among which in both cases is imagination, go to the making of a good artist or a good mathematician."

J. J. Sylvester (1814-1897), one of the greatest English mathematicians of the nineteenth century and the first exponent of higher mathematics in the United States, goes so far as to assign a definite place to mathematics as a fine art. He gives a geometrical picture of the mutual relations of the arts.

"Surely," he says, "the claim of mathematics to take

place among the liberal arts must now be admitted as fully made good. It seems to me that the whole of esthetics may be regarded as a scheme having four centers, which may be treated as the apices of a tetrahedron, namely Epic, Music, Plastic, and Mathematics. There will be found a common plane to every three of these, outside of which lies the fourth, and through every two may be drawn a common axis opposite to the axis passing through the other two. So far is demonstrable. I think it also possible that there is a center of gravity to each set of three, and that the lines joining each center to the outside apex will intersect in a common point—the center of gravity of the whole body of esthetic; but what that center is or must be I have not had time to think out."

A contemporary mathematician, D. Pedoe, reviewing a book on the "Foundations of Algebraic Geometry" says: "The ultimate aim of workers on the foundations of Algebraic geometries is to erect an esthetically pleasing structure, free from logical faults, on which the many ornaments of Italian geometry can be tastefully displayed."[3]

Julian L. Coolidge (1873-1954) concludes the preface to his "Treatise on Algebraic Plane Curves" with the forthright declaration: "The present author humbly confesses that, to him, geometry is nothing at all, if not a branch of art."

C · The Opinion of a Poet   Perhaps all this sounds like the preaching of professionals for the benefit of their own chapel, as the French proverb says. To allay such suspicions I could mention to the same effect the writings of Emerson, Thoreau, and others. I will limit myself to quoting a sonnet from the pen of one of the most talented contemporary American poets, Edna St. Vincent Millay.

SONNET

Euclid alone has looked on Beauty bare.
Let all who prate on Beauty hold their peace

And lay them prone upon the earth and cease
To ponder on themselves, the while they stare
At nothing, intricately drawn nowhere
In shape of shifting lineage; let geese
Gabble and hiss, but heroes seek release
From dusty bondage into luminous air.
O blinding hour, O holy, terrible day,
When first the shaft into his vision shone
Of light anatomized. Euclid alone
Has looked on Beauty bare. Fortunate they
Who, though once only and then from away,
Have heard her massive sandal set on stone.

D · Mathematics—a Creation of the Imagination   The objection may be made that, after all, these are the subjective reactions of some people, be they mathematicians or poets. If mathematics wants to lay claim to being an art, it must, in an objective way, show that it possesses and makes use of at least some of the elements which go to make up the things of beauty. What is the most essential element of an art? Is it not imagination, creative imagination? That mathematics exhibits abundantly this precious quality has been pointed out in the passage which I quoted a while ago. But it would be good to insist on this subject a little more.

Let us take any object of geometrical study, say, the circle. To the so-called "man in the street" this is a rim of a wheel, with, perhaps, spokes in it. Elementary geometry has crowded this simple figure with radii, diameters, chords, sectors, tangents, inscribed and circumscribed polygons, a transcendental ratio of the circumference to the diameter, and so on. By a further flight of the imagination it has wrapped around the circle all the lines and all the points of the plane in the fanciful theory of poles and polars. Here you have already an entire geometrical world created from a very rudimentary beginning. As if this were not enough, the mathematician has by a further effort discovered two imaginary points on every circle, the same two points on all the circles of the plane, and, for the sake of good measure, I suppose, he placed these two points on the unattainable boundary of the plane, and christened them "the circular points at infinity." This is undeniable proof of the creative power of the mathematician, for no mortal eye has ever beheld these points, or ever will. But as though to show you that nothing could stop him, the mathematician allows the whole circle to vanish, declares it to be "imaginary," and then keeps on toying with his new creation in much the same way and with much the same gusto as he did with the innocent little thing you allowed him to start out with. And all this, remember, please, is just elementary plane geometry.

Starting with ordinary integers, tools of common counting, the mathematician adds to them, in rapid succession, fractional numbers, irrational numbers, transcendental numbers. If these numbers still have something of the so-called "real" in them, the mathematical imagination can treat you to complex numbers, and if that be not enough, to ideal numbers. But with all that the mathematician finds that this is too hampering, too confining. He cuts loose from all numbers. Supremely confident in his creative genius, he declares, like an olympic god: "Let there be chaos and let me have a group of objects of any kind whatever in the chaotic world." Then he begins to promulgate laws by which these objects (or shall we say subjects?) shall be governed. He calls these laws relations, like reflexive, transitive, asymmetric, etc. Pretty soon he has what he calls a theory of aggregates. Further he creates such relations as "betweenness," "to the right," etc., and, lo and behold, here he has brought order out of chaos, and one of these orders is the very number system he used to amuse himself with, when the world was young. But there is no need for me to go to these abstract domains to see the mathematical imagination at work. It is enough to take cognizance of the immense literature massed around the elementary theory of numbers, which theory deals only with the ordinary integers, in order to get an idea of something which is incredible but true. Take the theory of groups, take the elusive, and yet enticing theory of probabilities, and so on. Truly, the creative imagination displayed by the mathematician has nowhere been exceeded, not even paralleled, and I would make bold to say, not even closely approached anywhere else.

E · Further Analogies Between Mathematics and the Arts: Symbolism, Condensation, Care in Execution, Etc. In many ways mathematics exhibits the same elements of beauty that are generally acknowledged to be the essence of poetry. Let us first consider a minor point. The poet arranges his writings on the page in verses. His poem first appeals to the eye before it reaches the ear or the mind. Similarly the mathematician lines up his formulas and his equations so that their form may make an esthetic impression.

Some mathematicians are given to this love of arranging and exhibiting their equations to a degree which borders on a fault. Trigonometry, a branch of elementary mathematics particularly rich in formulas, offers some curious groups of them, curious in their symmetry and their arrangement. Here is one such goup.

$$\text{Sin } (a+b) = \text{Sin } a \text{ Cos } b + \text{Cos } a \text{ Sin } b$$
$$\text{Cos } (a+b) = \text{Cos } a \text{ Cos } b - \text{Sin } a \text{ Sin } b$$
$$\text{Sin } (a-b) = \text{Sin } a \text{ Cos } b - \text{Cos } a \text{ Sin } b$$
$$\text{Cos } (a-b) = \text{Cos } a \text{ Cos } b + \text{Sin } a \text{ Sin } b$$

The superiority of poetry over other forms of verbal expression lies first in the symbolism of poetry, and secondly in its extreme condensation, in its studied economy of words. Take any of your favorite poems, or let us choose one of universally acknowledged merit, one we all know, say, Shelley's poem "to Night." Here is the second stanza:

Wrap thy form in a mantle gray,
    Star—inwrought.
Blind with thine hair the eyes of Day;
Kiss her until she be wearied out;
Then wander o'er city and sea and land,
Touching all with thine opiate wand—
    Come, long sought!

Taken literally all this is, of course, sheer nonsense, and nothing else. Night has no hair, night does not wear any clothes and night is not an illicit peddler of narcotics. But is there anybody balmy enough to take the words of the poet literally? The words here are only comparisons, only symbols. For the sake of condensation the poet omits to state that they are such and goes on to treat his symbols as realities. This economy of words is clearly apparent in the line "star inwrought." Just think of all the sentences you would have to string out, if you wanted to convey the same idea, and state explicitly all the similes involved.

The mathematician does things precisely as does the poet. Take the field of numbers. To begin with, the very idea of a number is an abstraction, a symbol. When you write the figure 3, you have created a symbol for a symbol, and when you say, in algebra, that a is a number, you have condensed all the symbols for all the numbers into one all-embracing symbol. When you further write a^m, the number m becomes a symbol once removed again. Then the mathematician becomes so confident in his symbols that he puts them to uses which he never contemplated and he often can not tell himself how far these symbols are removed from their original meaning. Think of the use of exponents as order of differentiation, or of derivatives of fractional orders, to take relatively simple cases.

As a symbol of a different domain, take the circular points at infinity referred to before. If you take the statements concerning them in a literal sense, they are obviously absurd.

But they are perfectly intelligible as symbols. It would take us too far afield to exhibit their precise meaning. Suffice it to say that they are a condensed, concentrated way of stating a long and complicated chain of rather simple geometrical relations.

This brings me back to the economy of words, a virtue the mathemetician has raised to the heights of a creed, against which no transgressions can be tolerated. He went so far as to do away with words altogether, or approximately so, and to replace them by special notations and symbols. This gives his statements a concision and an elegance quite inimitable. He has, however, to pay the penalty that his writings are incomprehensible to the non-initiate.

Another side by which mathematics approaches art is the care it exercises in regard to technique of execution. You do not enjoy a poem which is strained in the choice of words, where the rhymes are forced, a poem which bears on its face the marks of the labor of the poet. Not that the poet is expected to produce his poem without effort. Few poets do, and those only occasionally. We all know the stories about the pains poets of great renown have taken with their works, with each individual poem, with each line of the poem. But the result must be such that these labors are hidden behind an appearance of effortless ease. It is only then that you will grant that the poem is beautiful. The same is no less true of any other art. In the performance of music we go so far as to enjoy a piece of rather mediocre quality, if the piece presents considerable technical difficulties and the performer can show that they cause him no embarrassment.

Mathematicians are just as exacting with their technique of execution as any poet or artist is. They are constantly preoccupied with the elegance of their proofs or of the solutions of their problems. Any mathematician will immediately assign to the scrapheap any of his proofs, if he can think of another way to get the same result with less apparent effort, with the accent on the word "apparent." He does not hesitate to spend a great deal of extra time on the solutions he has already, if he has any inkling that he may abbreviate or simplify these solutions. And when he succeeds, when he has found this simplicity, he has the esthetic satisfaction of having brought forth an elegant solution. Nor is this effort limited to the individual. Mathematicians as a profession are always at work making the exposition of their science esthetically more satisfying. The success they achieve in this labor is often remarkable. Some of the results which the original discoverers have obtained in a most laborious way, making use of the most advanced and complicated branches of the science, may become, within a generation or two, very simple, very elegant, and based on almost elementary considerations. The beauty of this new way of execution becomes then the joy and the pride of the profession.

The methods of work of the mathematician and especially of the geometrician are very much like the methods of work of the poet. A physicist, a biologist, needs a laboratory to carry on his work. A painter needs brushes and paints, and canvas, and a studio as well; an architect is still more exacting. But the geometrician, like the poet, needs nothing at all for his work, beyond a scrap of paper and a pencil to help out his imagination by a rough and fragmentary sketch of the fleeting and complex creations he allows his fancy to play with. The geometrician, like the poet, is a dreamer, an incorrigible daydreamer. You may accuse both of them of absent-mindedness if you will, but neither of them would give up his daydreams for anything that the world could offer in exchange. These solitary dreams, these soaring flights of the excited imagination, make the geometrician, as they make the poet, oblivious to everything around him, make him forget his duties, his friends, his own self. But they are to him the most cherished happenings, the most precious moments of his life.

F · "Movements" in Mathematics   Mathematics resembles imaginative literature and the fine arts in general in that, taken historically, it has, like the arts, its "movements." During the second quarter of the seventeenth century analytic geometry came into being. Within a short space of time this "modern" geometry became the height of fashion in the mathematical world. The reputedly poised, sober-minded, matter-of-fact mathematicians became so infatuated with their newly acquired plaything, the Cartesian geometry, that they had no time and no use for anything that recalled the synthetic "method of the ancients." The splendid contributions of Pascal, Desargues, de la Hire, were consigned to accumulating dust for a century and a half.

This "movement" may perhaps be called the "symbolic" movement in geometry for it marks the period of introduction of algebraic symbols into the study of geometry. The turn between the eighteenth and the nineteenth centuries marks the beginning of another "movement" in geometry which may be called the "romantic" movement, for it corresponds, in time, to the romantic movement in literature. J. L. Coolidge, in an address before the American Mathematical Society, called a somewhat later phrase of the same movement the "heroic age" of geometry. The precursors of the movement were restored to their rightful place, and the enthusiasm for the new, projective, geometry was just as great and just as exclusive as in the case of the "symbolic" movement. Running true to form, the old school accused the new movement of mathematical heresy and denounced bitterly Poncelet's "principle of continuity."

G · Conclusion We have insisted on the common features of mathematics and poetry. One might think that we have overlooked the differences. Here is the way Prof. Buchanan, to whom I have already referred, formulates these differences. "The mathematician sees and deals in relations; the poet sees and deals in qualities. Mathematics is analytic, seeing wholes as systems of relations; poetry is synthetic, seeing wholes as simple qualities. The qualities that a poet sees are due to relations, says the mathematician. They need purgation. The relations that the mathematician sees are concrete and factual, says the poet. They need appreciation and love."

It is often said that mathematics should be studied for its usefulness. This is quite right. It should be studied for the philosophic insight it affords, and more so for the logic it uses and creates. But perhaps its main claim to your attention is based on the intrinsic beauty it reveals to those who can see it. The poet said: "Beauty is its own excuse for being." I would add to this: *The cultivation of beauty is its own reward.*

FOOTNOTES

[1] *Matematica y Poesia (essayo y entusiasmo)*, Arturo Aldunate Phillips, (Editiones Ercill, 1940).

[2] Those interested in this subject may consult:

Caskey, L. D., *Geometry of the Greek Vases.*

Ghyka, Matilla C., *Esthétique des proportions dans la nature et dans les arts.*

Hamridge, Jay, *Dynamic Symmetry.*

[3] Review by D. Pedoe, *Mathematical Gazette*, Vol. 31, 1947, p. 294.

Chapter V

MATHEMATICS AND THE MATHEMATICIAN

1 · IS MATHEMATICS AN EXACT SCIENCE?

A · *Mathematicians Are Human*

A · A Definition of Mathematics The famous eighteenth-century *Encyclopédie Méthodique* gave the following definition of mathematics: *C'est la science qui a pour objet les propriétés de la grandeur en tant qu'elle est calculable ou mesurable.* (It is the science which has for its object the properties of magnitude inasmuch as they are calculable or measurable.) Precise, concise, definite, and simple. This was in 1787. Even though this definition was adequate for the time, it was not destined to remain so very long. Two decades earlier Gaspar Monge (1746-1818) had invented descriptive geometry. He did not publish his results until 1795 because for over a quarter of a century the French High Command considered descriptive geometry its own private military secret. Monge's invention led his pupils to the creation of projective geometry, a branch of mathematics that does not deal with magnitude.[1] The quantitative conception of mathematics thus became obsolete. Many efforts have been made since to find a definition that would embrace all of mathematics. The enormous growth of the science during the past century and a half, and the inclusion of such branches as the theory of groups, topology, and symbolic logic, rendered all such attempts unsatisfactory. The hopeless task was finally given up in favor of simply saying that mathematics is what mathematicians are doing.

B · Is the Mathematician "Objective"?   How do mathematicians acquit themselves of the heavy responsibility that such a definition puts upon their shoulders? They have the advantage that they start out with a great amount of credit. To the layman mathematics is synonymous with exactness, nay, with certainty. Mathematics is precise, mathematics proves all the assertions it makes, all the propositions it advances. And books written by mathematicians seem to bear out the layman's opinion about the authors. These books seem to be written with complete detachment and strict objectivity. There is not a single exclamation point to be found on any of their pages, except when it is used as a symbol for a factorial. But do mathematicians actually do their work with that Olympic impartiality that the final product seems to exhibit?

Ask any mathematician worth his chalk why he spends so much time and effort on his research, and he will almost invariably tell you—quite truthfully so—that he does it because he finds it very interesting, because he loves to do it, because to him it is a most exciting adventure. Sentimental reasons all. But is sentiment a reliable partner of objectivity?

C · Priority Disputes   Every active mathematician will readily agree that he is trying by his efforts to promote and advance the science of his choice. There is no doubt that he is telling the truth and that he is quite sincere about it. But is it "the whole truth and nothing but the truth?" If it were, if the mathematician were interested in the promotion of his science in a purely objective way, it would make no difference to him whether it was A or B that took a given forward step, as long as the advance had been accomplished. But this is not the case, as is abundantly proved by the historically famous, and disgraceful, controversies over priority rights of mathematical inventions. The Newton-Leibniz quarrel over the invention of the calculus was just as bitter as it was harmful. It actually hindered the progress of the calculus in Britain for over a century.

The dispute between Poncelet and Gergonne as to who was the rightful owner of the title to the invention of the principle of duality may have yielded in scope to the Newton-Leibniz controversy, but it was fully as acrimonious, if not worse.[2] One could cite the quarrel between Descartes and Fermat, between A. M. Lengendre (1752-1833) and C. F. Gauss (1777-1855), and so on and on, *ad nauseam*. Cardan (1507-1576) obtained from Tartaglia (1500-1557) the solution of the cubic equation under oath of secrecy and then not only published the solution, but claimed it as his own. Our methods may not be as crude, but we are as jealous of our priority rights now as anybody ever was. Editors seem to think that priority claims are established by the date a given article reaches their desk, and publish this date as a part of the article. Perhaps whether it was A or B that made a contribution may not be of so much moment, but whether it was *I* or not *I* is of tremendous importance. The sublime indifference toward public acclaim exhibited by a Fermat does not seem to be of this planet. It may be argued that mathematicians as a rule get little else for their labors; they are therefore at least entitled to the honor and recognition their accomplishments can bring them. This is true enough. But it is a weak argument in favor of the supposed detachment and objectivity with which mathematicians view their work.

D · Withholding Results   There are cases on record when mathematicians were reluctant to publish the results of their findings, their reticence motivated by their solicitude for their science. When the researches of the Pythagoreans brought them face to face with irrational numbers, they were overwhelmed by their discovery. It contradicted the fundamental tenet of their philosophy that everything is (rational) number. The surest way out was to make of this troublesome result a professional secret and to induce the gods to destroy anyone who would dare to divulge to the lay crowd the exclusive wisdom with which only the initiates could be trusted.

We have a similar example in modern times. Gauss was in possession of non-Euclidean geometry ahead of both Lobačevskiï and Janos Bolyai (1802-1860), but he was loath to publish his results. He feared that such an unorthodox discovery might undermine the faith of the young in the validity of mathematics in general.

The judgment of both the Pythagoreans and of Gauss as to the effect of their discoveries upon the development of mathematics was totally wrong. But this is here quite beside the point. What is important to note in this connection is that the concealing of the truth is hardly the proper method to inspire confidence in the exactness of the science one is trying to promote.

E · Mistakes of Mathematicians   For men who are supposedly dealing with an exact science, the number of mistakes mathematicians make is both puzzling and disconcerting. The Belgian mathematician Maurice Lecat published a collection of *Erreurs des Mathématiciens*.[3] The list of names mentioned looks pretty much like a "Who's Who in Mathematics." Henri Poincaré was awarded a prize for a paper that had a serious mistake in it. He detected the error himself while his paper was in the process of being published, but it was too late to

remedy the situation, and the King of Sweden formally con-
ferred upon the author a prize for a paper that was wrong.

F · Disputes Over Results Obtained In an exact science it
should be easy to evaluate the merits of a paper, and experts
in the profession should be able to decide which of several
solutions of the same problem is the correct one. But this is
only too often not the case. Here is one example, of many that
could be quoted. In a paper *Fourier's series,* published by the
Mathematical Association of America, R. E. Langer (b. 1894)
relates a controversy participated in by d'Alembert (1717-
1893), Euler, and Daniel Bernoulli (1700-1782). Each of
these luminaries wrote a paper on the problem of vibrating
strings. The three-cornered polemic lasted more than a decade.
The only point of agreement that emerged clearly was that
there was always a two-to-one majority that the third party
was wrong. Human, all too human. But where does the exact
science come in?

B · Schools of Thought in Mathematics

G · The Quest for Rigor In spite of all these foibles, mathematicians mount a vigilant and jealous guard over the exactness of their science and are not a bit sparing of one another
when the impeccability of that science comes into question.

The invention of the calculus provoked a flood of criticism as to the mathematical and logical soundness of the new
doctrine. Neither Newton nor Leibniz was quite convinced that
the reproaches were groundless, but they found no way of
disposing of them.[4] Leonard Euler (1707-83), their most
distinguished immediate continuator, paid still less attention
to this controversy. He used his great gifts to expand and enrich the work of his illustrious mentors, and his unerring
instinct for what was right kept him firmly on the straight path.
However, Lagrange (1736-1813), a younger contemporary

of Euler, did not share the faith of the courtier of the czars of Russia in the formalism of mathematics. In Lagrange's estimation Euler's calculus "did not make sense."

The mathematical analysis bequeathed by the eighteenth century appeared to the mathematicians of the early nineteenth century to be a structure totally devoid of any foundation. Under the leadership of A. L. Cauchy (1789-1857) they undertook to provide analysis with underpinnings solid enough to render this branch of mathematics impervious to the most exacting criticism and at the same time to safeguard the results of mathematical analysis from all possible errors.

Thus came into being the school of rigor of the first half of the nineteenth century. It accomplished a great deal, but its achievements were anything but final. The second half of the nineteenth century set new goals for vigor. An attempt was made to "arithmetize" mathematical analysis. J. W. R. Dedekind (1831-1916) produced his theory of irrational numbers, Georg Cantor the theory of point sets, and so on. And the quest for rigor is still on the march. What satisfies the most rigid requirements of one generation of mathematicians seems totally inadequate for the next. E. H. Moore (1862-1932), for many years professor of analysis at the University of Chicago, expressed this in the apt adaptation of a biblical phrase: "Sufficient unto the day is the rigor thereof." It would seem, however, that mathematical rigor is a very elusive thing. The harder it is pursued, the more adroitly it evades the pursuer. In spite of all the advances that the nineteenth century contributed toward mathematical rigor, the mathematicians of the present generation feel that they are more "up in the air" than any other generation ever was.

H · Euclid and the "Obvious" Foundations of Mathematics
As a textbook Euclid's *Elements* has no rival, not only in mathematics, but in any other subject. More people over more centuries have learned their geometry from that book than have

learned any other subject from any other single book, with the exception of the Bible. And yet this is not the greatest of the merits of the book. The great role that this book played in the cultural history of mankind is due to the fact that Euclid's *Elements* was the first model of a deductive science. Euclid begins by defining the entities he is going to consider: point, line, angle, etc. Then he lines up his axioms and his postulates, i. e., those propositions that he accepts as valid on account of their plausibility or "obviousness." All the propositions that follow are derived from those assumed by pure reasoning, according to the strict precepts of logic. For some two thousand years there was nothing that approached Euclid's model in perfection.[5]

I · Formalism It is a queer irony of our intellectual history that it is precisely this perfection of Euclid's geometry that inspired the invention of non-Euclidean geometry. All through the ages students of geometry felt that Euclid's parallel postulate was not sufficiently obvious. Now a blemish on the perfect work of Euclid was an insufferable thing which had to be removed. The simplest and surest way to achieve this aim was to provide a proof for that postulate. But the many and various attempts to prove it failed. In the first half of the nineteenth century Lobačevskiǐ and Bolyai, following Euclid's model, each independently constructed a non-Euclidean geometry by assuming that Euclid's parallel postulate is *not valid*. Each of them pushed his geometry far enough ahead to convince the most skeptical that their systems are quite coherent and not likely to run into inconsistencies. All doubt on this score was finally dispelled when it was shown that the Lobačevskian plane non-Euclidean geometry may be interpreted as Euclidean geometry on a pseudo-sphere.

The non-Euclidean geometries rendered Euclid's parallel postulate, if anything, even less obvious. Still Euclid succeeded in constructing his elements in spite of this deficiency. From this there was only one step to the conclusion that the logical coherence of Euclid's *Elements* is in no wise dependent upon the obviousness of its postulates, and that it should be possible to build a consistent geometry with a set of postulates that would lay no claim to obviousness whatever.

The basic entities of Euclid's great work fared no better than his axioms. It all started with the "principle of duality," to which allusion has already been made. This principle asserts that if in any valid proposition of plane projective geometry the words "point" and "line" are interchanged, the resulting proposition is also valid. This astounding discovery inevitably led to a strange conclusion, namely, that the nature of the basic entities to which the basic postulates of a deductive science are applied is quite immaterial. In fact, these entities need not have any meaning of their own. Their relation to each other is determined by the postulates that are applied to them, and that relation is all that matters.

On these foundations was built the "formalist school" of mathematics, of which David Hilbert (1862-1943) was the leading exponent, the high priest of the cult. There was, however, a bothersome fly in the ointment. In fact there were two such flies. If postulates for a mathematical science, for example, geometry, are set down arbitrarily, and if the entities to which they are applied are devoid of meaning, what relation does such a geometry bear to the physical world? Richard Courant (1888-), a former colleague of Hilbert, says in the preface to his *"What is Mathematics?"*[6] that such a doctrine "is a serious threat to the very life of science," that "such Mathematics could not attract any intelligent person." The formalists, however, made short shrift of objections of this kind as long as they could feel that their science remained logically without a blemish. On that ground they were undeniably right. But it was not so easy to kill the other fly, for nothing less was involved there than the logical foundation of the formalist science.[7]

The "obviousness" of Euclid's basic propositions referred to the fact that these propositions are extracted from our daily experience and are realized, somewhat crudely, in the world that surrounds us: they are thus consistent with one another. If the postulates are taken arbitrarily, if they have no intuitive connotation, what guaranty is there that they are logically consistent? Without a proof of the consistency of the postulates the whole edifice is worthless. The formalists realized that no less than their bitterest critics. Hilbert made heroic efforts to find such a proof. He failed. And there the matter rests, except that it has been proved to the satisfaction of those most competent to judge that, within the framework of a given formalist science, it is not possible to find a proof that science is consistent. If a proof of consistency for a formalist science is to be produced, it must come from outside that science. This proposition is due to K. Goedel.[8]

J · Logicalism   The formalist school of thought in mathematics takes logic for granted. To this logic it adds an arbitrary set of entities—"undefined terms"[9] and an arbitrary set of postulates—"unproved propositions." It is then in possession of all the necessary tools and materials for the building of the proposed branch of mathematics.

Another school of thought came to the conclusion that the formalists are extravagant: they require too much. Logic alone is perfectly sufficient for the erection of the entire edifice of mathematics. Not the old verbal logic, but logic reduced to a set of symbols, after the manner of algebra. By means of this "symbolic logic," to give it its proper name, all mathematical entities, including the integers themselves, can be obtained by purely logical constructions. This philosophy of mathematics culminated in the three-volume work *Principia Mathematica* (1910-13) by A. N. Whitehead and Bertrand Russell. This was an extremely ambitious undertaking, undoubtedly one of the greatest intellectual enterprises of all time. It was hailed with great enthusiasm in England and in the United States. Helping hands came forward to render the great work still greater.

But the *Principia* began to suffer from the same malaise as Cantor's theory of point sets, as the infinite processes put to work to provide a logical foundation for the mathematical continuum. Paradoxes and antinomies came to light that were very embarrassing. Some of the fundamental assumptions of the *Principia* introduced for the express purpose of warding off paradoxes were found to be questionable and finally rejected. It was not long before the *Principia Mathematica* was reduced to the status of one more contender for the honor of being the custodian of the foundations of mathematics, under the name of "logicalism."

K · Intuitionism   Among the critics of the *Principia* were the French intuitionists: E. Borel (1871-1956), Lebesgue, and others. But the greatest challenge of this work came from members of the Dutch school, called by Abraham A. Fraenkel the "Neointuitionists." This school, under the leadership of L. E. J. Brouwer (1882-), put the *Principia* upside down. Not only did they reject the idea that mathematics can be derived from logic, they denied logic any autonomous existence. Logic, according to the intuitionists, is not a science but a technique derived from science to facilitate the study of the science. Furthermore, Brouwer boldly questions the validity of the basic processes of our generally accepted logic. He rejects the law of the excluded middle, i.e., that a proposition is necessarily either true or not true. It may be neither, for there may be no sufficient information to decide the question.

As an illustration of what is meant by Brouwer's negation of the law of the excluded middle, let us consider the example given by Abraham A. Fraenkel.[10] The fractional part of the number π has been computed for many hundreds of places, and many more such places could now be computed with much less labor than before, by means of the new electrical calculators. Is there a place in this long row of numbers where the digit 7 occurs seven times in a row? There is no such place in that part of the fraction that is known at present, and we cannot tell whether it will or not occur if new digits of that fraction are computed.

Now let us consider the real number R which starts out as 0.333333 and every other digit of this decimal fraction is a 3, except that if the nth digit of the fractional part of π is a 7 followed by six more digits 7, we will take for the nth digit of R the digit 2, if n is odd, and the digit 4, if n is even. The digits of R are thus perfectly defined as far as the digits of the fractional part of π are known. But we cannot tell whether R is equal to $\frac{1}{3}$, smaller than $\frac{1}{3}$, or greater than $\frac{1}{3}$.

Is the famous saying "You cannot fool all the people all the time" true or false? Perhaps it is true. But it is conceivable than a man publicly perpetrated a hoax or a lie that remained undetected during his lifetime and that he took his secret with him into his grave. Then the proposition would, of course, be wrong, but we would have no way of proving it. If the man wrote a confession, sealed it, and ordered his heirs to open it on the one-hundredth anniversary of his death, then we shall find out on that day that our proposition is false. But at present the proposition is neither true nor false. Hitler was quite certain that the proposition is false. Witness his principle of "the big lie."

"Francis Bacon (1561-1626) is the author of the so-called Shakespearean plays." Is the proposition true or false?

L · New Logics Things did not become any smoother for any of the contending schools of thought when the Polish logician Lukasiewicz raised the question why logic should be limited to only two alternatives, two values: true and false. He proposed a new logic which admits of three alternatives, a three-valued logic. Now *ce n'est que le premier pas qui coûte.*

If logic can be three-valued, why can it not be four-valued, indeed, why not n-valued? There is no reason, however, to stop there. Why must the values of logic be a finite whole number? We might as well have a logic with a continuous number of values—such proposals have been advanced.

That unshakably solid rock of classical logic simply slipped away from under the mathematical edifice, and the whole structure is now "on the rocks." As mathematicians put it, their science is at present passing through a "crisis." It has been in this state, roughly, since the beginning of the present century. What connection, if any, is there between this crisis, and the social and political turmoil in the throes of which suffering mankind has been laboring during the same period of time? This is not the time nor the place to consider this question, but so far as mathematics is concerned, one need not be overly alarmed. Mathematics is not going to the dogs.

M · Conclusion   Mathematics has two aspects: On the one hand, it is a description of a segment of the world we live in and it furnishes tools for non-mathematicians to describe other segments of that world. This might be called the "functional" part of mathematics. The other part of mathematics deals with its foundations and may be said to be largely philosophical. Of course the two parts are not unrelated. The study of the foundations of mathematics decides how far the mathematical processes may be carried out before they reach the limits of their validity. Fortunately, whatever these limits may be, there is ample room for mathematical activity long before those limits are reached. As a mater of fact, most active mathematicians are little concerned about those foundations. At least they do not allow those problems to interfere with their activities as investigators. More than that, even those mathematicians who take a direct part in the debate regarding the logical validity of their science manage to obtain very valuable results in their own special field of investigation that have little relation to those theoretical discussions.

But what about the crisis itself? It would, of course, be foolhardy for anyone to try to predict at present where the crisis leads to and how it will end. What may be said, however, with perfect safety is that mathematics will emerge from it enriched and invigorated, to continue the work it has been so successfully carrying on up to now.

2 · PERPLEXITIES OF A POTATO-PUSHER

A · Winning a Prize   The peace of mind of the reader may perhaps have been disturbed by this title, for "potato-pusher" is not in the dictionary, not yet. If you are puzzled as to what a potato-pusher might be, I must hasten to put you at ease, by supplying the aforesaid deficiency. Unlike a potato-peeler or a potato-masher, a potato-pusher is not a kitchen utensil but a person, and in the present circumstances the reference is to no other but myself.

I am quite sure that all of you will agree that I ought to be preplexed, for many more reasons than one. But some of you might wonder on what ground I arrogate to myself the high-sounding title of a potato-pusher. Those benighted individuals have only themselves to blame for their ignorance. They should have attended the party given by the Department of Mathematics and Astronomy some time ago in the Faculty Club. Had they been there, they would have witnessed, they would have seen with their own eyes the prowess I displayed then and there as a potato-pusher. Why, I was the champion of the evening and won the prize, the only prize, mind you, that was awarded. When all the nice ribbons were untied and all the multitudinous pretty wrappings undone, there was the prize, for everyone to see: twelve round, neatly packed, nice little potatoes. I am afraid that some echoes of a malicious whispering campaign reached your ears that mine was the booby prize. I am sure of that campaign, for I heard it myself, all the way across, from the other end of the room. But never you mind. You know how some people are: envious, always ready to belittle a fellow, to deprive him of his just credit, of his hard won dues. I am the champion potato-pusher whether they like it or not.

B · Gambling and Statistics   But I must admit that the prize did not do me much good. For it set me a-thinking, and as you know thinking is a weariness of the flesh. The more I thought, the more worried, the more perplexed I became. Not that there is anything wrong in winning a prize, from any point of view, least of all from a mathematical point of view, as I could readily prove to you by any number of examples. Let me just tell you one story, an excellent story, even if it is a little better than three centuries old.

Chevalier de Méré was both a nobleman and a gambler. He had the good fortune to count Blaise Pascal (1623-1662) among his friends. The noble gambler once asked his erudite and resourceful friend to suggest a fair way out of a difficulty in which he was involved. To put the story on an impersonal basis, let us say that two players A and B, of equal skill, agree to play a game for a prize which is to go to the player who first wins three games. When A had two games to his credit and B one game, the contest had to be given up. What would be an equitable way of dividing the prize between A and B?

Pascal communicated this question to Fermat, and between them the two mathematical geniuses of the first half of the seventeenth century evolved two solutions of the problem which were just as simple as they were ingenious. Let us suppose that A and B play one more game and that B wins it. With two games to the credit of each player, they should divide the prize equally, which is to say that half the prize certainly belongs to A right now, before the hypothetical next game is played, and that game is played only to decide what to do with the other half. Now A has as much of a chance to win that game as B does, hence that second half should be divided between them equally. Thus one-fourth of the prize should go to B and three-fourths to A.

The second solution is even simpler than this. B can win the prize only if he wins two games in succession, that is to say, he has one chance in four, like throwing heads with a coin twice in a row. Hence one-fourth of the prize should go to him and the rest to A.

Little did the Chevalier suspect that the trend of thought provoked by his prize problem would lay the foundation of the theory of probalities, and that this, in turn, would lead to the mighty discipline now known under the name of Mathematical Statistics.[11] The practitioners of this new craft are so proud of their calling that they scorn the title of Mathematicians who are engaged in the study of statistics. They insist that they are Statisticians who use mathematics as a tool, say, like physicists, or engineers. In evidence whereof they formed their own Statistical Society, separate from the Mathematical Society, and have their own exclusive Quarterly of Mathematical Statistics. This shows that the Statisticians are even prouder than the topologists. But far be it from me to be putting ideas into the heads of the topologists.

C · Tit-tat-toe Ancient and Modern   All this mighty development came about because of Chevalier de Méré's prize. But my potato prize brought me nothing but perplexities. This potato-pushing reminded me of other games I used to play at one time or another. The first that came to my mind is one that some of you may know under the name of tit-tat-toe and I knew under an entirely different name. The equipment necessary for the game consists of a square divided into nine smaller equal squares, or cells, and two sets of three chips each. The two opponents move their chips in turn, one at a time, and the one who places his three chips in a horizontal or vertical row is the winner.

I used to play that fascinating game when I was in the grades. My favorite time for the game was during school hours, especially during the arithmetic lessons, when the subject became too repetitious and too boring. In the school I attended the pupils had no individual desks. We were seated on long benches, like church benches. I had no trouble in inducing a neighbor of mine to play the game with me. I used the simple device of bribing him with the promise to show him my solution of the next day's assignment. We manufactured the necessary equipment right on the spot. Two pairs of mutually perpendicular lines drawn on a scrap of paper served as the board, and the chips were six bits of paper, three marked with rings, and the other three with bars. We played to our hearts' content and had the time of our life.

I have quit playing tit-tat-toe a long, long time ago, and I am glad I did. For I have found out that mine was "child's play." Self-respecting people with proper mental equipment do not play the game the way I used to. For poise and dignity the game is to be played in three dimensions. The "board" of the game is a cube sub-divided into twenty-seven smaller and equal cubes, or cells. Of course, I knew nothing of all that in my tit-tat-toe days. Besides, what good could that have done me, had I known it? You would agree that such a device could hardly have escaped the benevolent and vigilant eye of my teacher, and my perplexities would have started right then. Besides, even this high-brow style of playing tit-tat-toe is obsolete, just as obsolete as the carriage of King Tut-ankh-amen. You see, two mathematicians have gotten hold of that game recently. They freed this pastime of all triviality and endowed it with the proper intellectual prestige by elevating it to the fourth dimension, nay, to the nth dimension. Yes, if you want to keep your self-respect and keep up with the times, you must play your tit-tat-toe at least in the fourth dimension.[12] So far I have not played this hyperspatial tit-tat-toe. Why? The reason is very simple. Nobody has as yet tried to bribe me into playing four dimensional tit-tat-toe by offering to solve my problems for me. I mean my mathematical problems—my other problems, and especially my financial problems, I know for certain to be insoluble—like the problem of the duplication of the cube, or the solution of the nth degree equation.

D · New Checker Games for Old My triumphant exploit in potato-pushing made me also think of my checker days, that is, of the days when I used to play checkers. Chronologically that was after my tit-tat-toe days. But I gave that game up, too. Roughly speaking, that happened when my mathematical problems became tough enough and challenging enough so that I became satisfied to grapple with them all by myself in the solitude and the silence of my study. I no longer felt the need of the stimulus which is provided by the opportunity to gloat over the demise of a defeated opponent, or, what is the same thing, the stimulus provided by the gratification of my ego in feeling superior to some one else. But am I right about that? Now that I said it, I am afraid that, upon second thought, I may have to take it all back. What about that nasty fellow with his mocking grin on his repulsive face who always peeps over your shoulder at every word you put down on paper, instantly ready to jump on you with his priority claims of having "got there firstest, with the mostest and the bestest arguments?" (With apologies to General Bedford Forrest, of Civil War fame). I really love to beat that guy to the punch. Don't you?

But it matters little what the actual motive was that made me abandon checkers. The point is, I am glad I did. For I found out that this, too, is an antiquated game, at least in the form I used to play it. For one thing, there is no good reason why the game of checkers must be played so that the typical move of the typical piece must always be in a straight line. The game could be played with two-dimensional moves. That would have the immeasurable and enticing advantage that you could move across an edge of the cell, or through a corner, or both, if you are very ambitious. Just fancy how much checker liberty would be yours, to hold and to cherish. However, with all these up-to-the-minute improvements in your checkers, you would still be playing a piker's game. The real thing is to play checkers in three dimensions.

Using a suitable frame, several excellent cube checker games can be defined, with many interesting new features. The field of play may be a network of white and black cells, or a looser network of cells holding together by their corners, or the entire frame. Places of local safety, like the familiar double corner, and other strategical features, appear in new forms. There are many possible kinds of cube checker games, pure, combination, and hybrid games, multiple games, interfering games, cyclical games, and others. The best have already proved more interesting than the classical checker game.

I hope this does not make you feel dizzy. If it does, don't blame me. I have not invented it. Nor am I reporting a kind of fly-by-night scheme. I heard this three-dimensional checker game expounded under the auspices of the American Mathematical Society at its meeting at Cornell University.[13] And the American Mathematical Society, I want you to know, is the largest, the richest, and most powerful, the most influential, and the most authoritative organization of mathematical research workers in the world today.

However, if you still have a weakness for the traditional, though outmoded checker game, you may still hope for a respite for some time to come. The exposition of the three-dimensional checker theory was illustrated on an actual model. My innate simplemindedness pushed me to ask the very naïve question where such a progressive and up-to-the-minute outfit could be secured. I was promptly put in my place by the declaration of the speaker that as far as he knows the model before him is the only one in existence at the present time. So the flatwitted checkers will continue to flourish for some time. But the millennium of progress is at hand.

You need not fear that the tit-tat-toe game has anything on checkers. The speaker was magnanimous to assure his breathless audience that checkers, too, could be played in dimensions higher than the third. The anxiety of all present was visibly relieved. But, believe it or not, this assertion was not accompanied by the exhibition of a model of such a game in n dimensions. No explanation for this omission was offered. That this was a grievous oversight was quite clear to me right on the spot, but I dared not ask any more questions.

E · Potato-pushing à la mode By now, I am sure, you realize already why the aftermath of winning a prize turned out to be so full of perplexities for me. It suddenly dawned upon me that the time is ripe to generalize the potato-pushing game. And who is to do it, if not the champion. What a wonderful opportunity! What an alluring vista! The portals of immortality have suddenly swung wide open right in front of me, beckoning me to enter and join the illustrious and enduring company of generalizers which dwells within. Small wonder that the prospect turned my head. I also realized that I have no time to waste, for I must make sure and run past that gate "firstest." Yes, but how do you generalize a potato-pushing game?

I understand that the present form of the game is already the result of some evolution. In a preceding stage the instrument with which the potato was pushed was not a stick, but a part of the player's anatomy, like the nose. Is that the avenue of approach? Clearly, that would be retrogression. Progress does not point in this direction. Besides, I do not think I would particularly enjoy the game if I had to push the potato, say, with my tongue in my cheek.

I hoped that some solution will be suggested to me by your woman's intuition. I do not mean my wife's intuition, I mean my own. But nothing of the sort came to relieve me of my perplexities. I was therefore reduced to the slow and laborious method of analyzing the problem in detail, trusting that such a systematic procedure may yield some salutary ideas. How can the game of pushing a potato with a stick be described in general and scientific language? Suppose I say that the game may ideally be conceived as consisting of the propulsion, over a plane, of a ellipsoid of revolution by a straight line, along a prescribed path which is also a straight line. If this is an acceptable way of looking at the thing, I am about to see a glimmer of light. I could give up the prosaic, down to the earth straight line and make the potato move along, say, a spiral of Archimedes instead. You would admit that my spiral, or rather Archimedes' spiral, would make the game much fancier, would it not?

But I could not be satisfied with that. The game involves four geometrical elements, and there is no good reason why one of them should be singled out, and the others neglected. This would be contrary to the democratic spirit of the times. So my perplexities would continue, until I generalized each element in various ways and evolved a great many combinations, too numerous to mention. To give you an idea of what they were like, I would say that we could roll a pseudosphere along a geodesic curve of a Frenel wave surface and use a witch of Agnesi to propel it. I felt quite certain that if I offered to the world a few dozen of such improved potato-pushing games, I would contribute powerfully to the joy and happiness of mankind and earn thereby all the acclaim and all the gratitude a man can wish for.

But that blissful state of mind lasted only a short while. My perplexities returned to plague me some more. I noticed that all four elements of my game, much improved as they were, have kept their original dimensions. That makes me a very poor generalizer, and I am much more likely to be laughed at than commended. And rightly so. Why such conservatism? There remained the question, which of the four elements involved shall undergo a change of dimensions? To make a long and painful story short, I will tell you that I came to the inescapable conclusion that in order to do the thing properly I must strike out boldly, take the bull by the horns, and go the whole hog. I must assign to the four elements involved the dimensions p, q, r, s. To make sure not to be outdone by anybody, I must allow p, q, r, s to be any four numbers whatever, positive, negative, fractional, transcendental. Now I have it. "Eureka." I rested on my well-earned laurels.

But not for long. There is no rest for potato-pushers, or rather generalizers. All of a sudden I realized that in my thinking I had been visualizing Euclidean space. Such a limitation is absolutely intolerable. It is imperative that non-Euclidean spaces be brought in, and non-Archimedean spaces, and non-Arguesian spaces, too; and some purely topological spaces, like the Banach space, must not be neglected either. With such improvements our potato-pushing game will defy all competition.

But this feeling of having reached the ultimate did not last. The game involves motion. Now motion is relative. How is one to tell which element of the game is to remain stationary and which is to move? To come back to the almost forgotten prototype of our game, why is it necessary to move the potato over the floor, when the same result could be obtained if the potato were kept fixed and we pushed the floor about. In the generalized game the same argument may be applied to any one of the four elements involved. If we carry this idea out to its logical limits, what a wonderful game we would have. But where am I? Do you know?

You may think that by now the game is general enough. But this is not so, not if you are a potato-pusher worth your salt. My perplexities and my worries were back, worse than ever. At this stage it occurred to me that there is a shortcoming that is common to all the generalizations I am familiar with. It was quite evident that I cannot afford to be caught in this kind of treacherous trap myself. I have reference to the patently notorious fact that all the generalizers have overlooked the player himself. They left him invariant. This is unworthy of an honest to goodness generalizer, let alone a potato-pusher. This idea is not quite original with me. If mathematicians never thought of it, thieves have actually practiced it since ancient times. You know of the famous robber of ancient Greece by the name of Procrustes who made his victims fit the length of the bed he kept in readiness for them, either by stretching them, or shortening them, with an axe if need be. We might say that mathematically speaking, Procrustes submitted those whom he robbed to a linear transformation. In all fairness this transformation should be called a "Procrustean Transformation." When it comes to generalizing games, mathematicians should look for inspiration to the robbers of antiquity. The players of the game should be subjected to a Procrustean transformation which, of course, does not necessarily have to be linear. The exigencies of the game under consideration would decide that question of the details of transformation to be used. In the particular case of the potato-pushing game I wonder whether the purposes of the game would be better served if my Procrustean transformation should reduce the player to two dimensions, or on the contrary he should be blown up by that transformation to four dimensions, or even higher dimensions. This is still one of my unresolved perplexities.

F · Conclusion That is as far as I got. Have I done everything that can honestly and properly be expected in the responsible task of generalizing the potato-pushing game? I do not know and I am perplexed. You see, after having played his first and only game of potato-pushing a fellow, although a champion, does not have the proper perspective nor does he have the requisite insight to do justice to the game in the way of improving and generalizing it. I may or may not gain immortality in the attempt, but I am sure that I, too, can generalize all the fun and all the joy out of the game, and every speck of common sense along with it. You just give me a chance.

3 · GEOMETRICAL MAGIC

A · A Point Fixation Rather inadvertently I found myself not long ago in a quite sophisticated gathering. The company was being entertained with a variety of tricks by a skillful magician. For most of the numbers of the puzzling show the performer was enlisting the active participation of some members of his audience—the most successful of his stunts, I should judge.

At one point of the spectacle the magician issued a call for a new kind of help: "Would someone assist by drawing parallel lines?" An uneasy silence fell abruptly upon the amused crowd. Nobody budged, for what seemed a long time. To save the situation from becoming too embarrassing I volunteered, foolhardily. The magician sized me up with a disapproving eye, but he said nothing. The poor man had no better choice.

I was armed for my task with two triangular rulers, in addition to the pencil, and confronted with a large triangle ABC drawn on a sheet of paper. The audience eagerly crowded around the big table, as though expecting I do not know what miracle. The magician planted himself right next to my chair to direct operations. The choice of the starting point, say X, on the base BC of the triangle was mine, but the rest of the procedure was strictly prescribed. First I had to draw a line XY parallel to AB and terminated on AC at Y. Then I was bid to draw a parallel YZ to BC reaching AB in Z, and finally a parallel ZX' to AC, thus returning to the base BC, at the point X'. (Fig. 10)

I was quite pleased with my feat, and I was glad it was over. That feeling of relief, however, was short lived. The magician dared me to draw a second sequence of three lines, analogous to the first, but starting this time with my hard-won point X'. I could not think of any good reason why I should decline, the more so that I felt safe enough, since I knew what was ahead. To avoid an argument, I bravely drew the lines X'Y', Y'Z', and was ready to mark triumphantly my terminal point X'' on BC, when suddenly, and almost involuntarily, I jerked my pencil away from the paper: my point X'' fell so dangerously close to the initial point X that it was hardly possible to tell them apart.

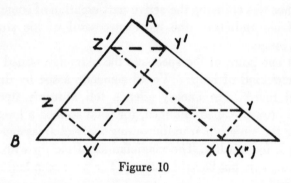

Figure 10

"My parallels are not much good," I said sheepishly, looking up to my mentor, or better, tormentor.

"On the contrary, sir," he said consolingly, "your parallels are amazingly exact."

"Then how is it that I landed in occupied territory? Is there anything wrong with the starting point I picked?"

"You would not want me to surmise," he said slyly, with perceptible mockery in his voice, "that you have a bad conscience about that point and feel impelled to return to the place of your original sin."

"But to be frank with you," he added after a barely noticeable pause, "you could not help coming back to that point, sin or no sin. It is the effect of the magic spell you are under right now. It is a fixation with you."

I understood quite well that this speech was meant primarily for the benefit of the onlookers who seemed to be quite intrigued by my drawing, or perhaps simply enjoyed my obvious discomfiture. Be that as it may, the choice between a guilty conscience and a magic spell to justify a geometrical figure was for me too harrowing, too upsetting an experience. The magician seemed to have sensed that by his magic, I presume, or he may have read it in my face. For after a little while he addressed me again, in a most condescending manner.

"If you are troubled in your mind, sir, by the double dose of parallel lines I imposed upon you, I shall show you my magic powers over you in a simpler way. You may still choose the starting point X at will and then proceed to draw only the first set of three parallel lines. But before you draw those lines I will mark the point X' on BC where you will land."

By that time I felt that I had nothing more to lose in the way of dignity, or prestige, and I might perhaps gain some insight into the trickery of the magician. I accepted the offer. I tried the thing twice, once taking X on BC, and the other time taking X on BC extended. I drew my parallels very carefully, but no matter how hard I tried, the confounded magician came out on top: I ended up each time mighty close to the spot he designated beforehand.

I could get nothing more out of my magician. I left the place feeling a dire need for some more light. Magic spell— piffle. But what else was there behind his uncanny ability to foretell where my parallels would lead me? It would make a tedious story if I attempted to tell you of all the schemes I resorted to in my efforts to break open that irritating puzzle. Suffice it to say that I was definitely determined to find an answer, and find one I did.

While I was proud and happy to succeed in tearing to shreds the veil of the magician's secret, there really is not much to crow about now, as I look back on it. You just notice that the figure as drawn includes two parallelograms BXYZ and CX'ZY, so that we have

$$BX = YZ = CX'.$$

If you bear this in mind it is perfectly easy to write out the following equalities

$$BX' = BC - CX' = BC - BX = CX.$$

There is nothing more to the "baffling mystery" of the magician than that $BX' = CX$. When he saw how far from C my point X was, he marked his point X' at the same distance from B. Or, to put it in other words, the points X, X' are equidistant from the mid-point A' of the base BC. In this light I see now that I could have put a troublesome crimp into the works of the magician, had I chosen for my point X the mid-point A', of BC as I had learned in my plane geometry. I wish I could have thought of it then!

What is still more curious, the observation that X, X' are symmetric with respect to A' dispels also the "magic" of the double set of parallels. Indeed, if I start with the point X' and draw the additional set of parallels X'Y', . . . I have to end up, according to that observation, at a point on the other side of the mid-point A' of BC, at a distance equal to X'A', and this is precisely the point X. Perfectly wonderful, is it not, and in spite of its simplicity or perhaps just on account of it?

Surely, there is no room and no excuse for reading magic into such an innocent figure. Yet I am somehow inclined to commend the magician for making this piece a part of his program. I am ready to agree that there is more "wondrous magic" in this one geometrical problem than in all the rest of the trickster's repertoire. But you have to learn to appreciate its enchantment, I guess, just as one has to learn to appreciate the taste of coffee, or the smoking of a pipe.

B · A Square Deal   Among my acquaintances of more recent vintage there is a man with a reputation of being something of a "big-shot" as a mathematician. Let us call him Null, for short. The other day, while visiting him, I could not resist the temptation to spring on him the problem of my magician.[14] I was put out quite a bit when Null claimed to be familiar with the problem and to know the proof thereof. But when I told him how I came into possession of this piece of mathematical learning, there was a look of surprise in his wide open eyes. "That's curious," he remarked calmly, "I was not aware that magicians make use of geometrical problems in plying their tricky trade." "But when I come to think of it," he added after a brief pause, "there is no reason why they should not. Many a problem in geometry, and in other branches of Mathematics, for that matter, may easily be dressed up so as to serve their puzzling purposes."

Now it was my turn to be surprised. "Could you perchance think of an example?" I asked somewhat dubiously.

"Well," Null said slowly, obviously playing for time, "to furnish such an example it is necessary to think of familiar things in an unusual way, in a way different from the customary one."

He got up, paced his study forth and back several times, evidently preoccupied. Suddenly he stopped right in front of me: "Would you be willing to act as my "medium" too, now that you are an experienced hand in the business?"

"I shall be glad to be of help, if I can," I said as casually as I could, ignoring his irony.

"This time the trick will be quite different: you will have to draw perpendiculars," he joked with growing good humor.

Under Null's direction, I marked four points A, P, Q, R in a row, entirely of my own choice, with the sole restriction that the segments AP and QR had to be equal. At the points P, Q, R, we erected perpendiculars to the line APQR and on them we marked the points B, C, D, so that PB = AQ,

QD=QR, RC=PQ, taking care that the points B, C be on the same side of the line APQR, and the point D on the opposite side of it. (Fig. 11)

"Now we are ready for the kill, I mean for the finishing touch," Null announced, evidently satisfied with my handiwork. "Let's join A to B and D, and again C to B and D."

I must have looked quite puzzled contemplating the completed drawing, for Null asked me with ill concealed amusement: "What's wrong?"

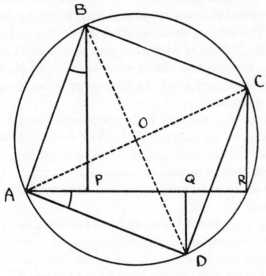

Figure 11

"It just looks to me that I have unwittingly drawn a square. In fact, I am ready to swear upon the beard of Pythagoras himself, that ABCD is as good a square as I have ever drawn before. It could not be an accident, could it?"

With strained reserve Null replied: "I would rather that you find the answer to your question yourself, if you do not mind."

I had to grant, in my own mind, that his was a reasonable attitude, the more so that it suited me quite well. So I went to work. I observed that the only element of latitude in the data of the problem was the spacing of the four points on the line. So I varied them all I could. But the ubiquitous square was there every time, in all its provocative and challenging exactness.

"Of course," Null commented when I declared myself satisfied as to the inevitability of the square, "you knew before hand, that this would be the case. Otherwise my game would have been pointless."

"Yes, Mr. Null, I realized all along that I could not win," was my humble admission, "but the drawing had a good effect upon my peace of mind, the professionals' prejudice against graphical proofs not withstanding."

"Your unprofessional behavior is far more excusable than my own," intervened Null in good humor, " and I transgressed in more ways than one. In the first place, professional custom, not to say professional ethics, would require that I tell you what the outcome of the chain of operations will be, leaving open the question as to the proof of that statement."

"And the temptation to produce a more telling effect made you deviate from the narrow path of virtue," I interposed with marked irony.

"Quite so, quite so." Null laughed. "But also because I worked under pressure. Your magician did not tell you what the climax of the performance would be. I had to meet the competition, had I not?"

"The statement of this problem is worded poorly," Null continued without waiting for me to grant him his pardon. "Moreover, not all the conditions mentioned are necessary to obtain the square. The problem could be stated as follows: Construct a square (ABCD) given one vertex (A) and the projections (P, Q) of the two adjacent vertices (B, D) upon a line (APQ) passing through the given vertex (A)."[15]

"This is a much more transparent way of putting it," I gladly conceded, "why then resort to the other?"

"Professional mathematicians are human, too, believe it or not," Null replied smiling indulgently. "They are not always out to enlighten their confrères. Sometimes they would rather dazzle and puzzle their brothers-in-arms. So they camouflage the source of their information the best they can and invite or challenge the others to try to unearth their secrets."

"I read somewhere," I said trying to parallel his argument, "that a fox walking on snow-covered ground uses his tail to obliterate his footprints, in order to keep the hunter from getting on his trail. Mathematicians sometimes try to be foxy, too."

"Now, now, my young friend, I am afraid you are waxing somewhat cynical," Null objected, his smiling eyes contradicting the mildly reproachful tone of his voice.

FOOTNOTES

[1] Cf. Chapter III, Section 2b.

[2] Cf. Chapter I, Section 4.

[3] *Erreurs des Mathématiciens*, Maurice Lecat (Brussels, 1935).

[4] Cf. Chapter III, Section 3e.

[5] Cf. Chapter I, Section 1d.

[6] *What Is Mathematics?*, by Richard Courant and Herbert Robbins (1941).

[7] Cf. Chapter I, Section 1d; Section 3f; Section 4b.

[8] *Ibid.*, Section 4b.

[9] Cf. Chapter I, Section 3f; Chapter III, Section 3f.

[10] *Scripta Mathematica*, 13, Nos. 1-2, 1947.

[11] See Chapter II, Section 1d.

[12] Funkenbush, William, and Eagle, Edwin, "Hyper-Spatial Tit-Tat-Toe" or Tit-Tat-Toe in Four Dimension," *National Mathematics Magazine*, Vol. 20, No. 3, December 1944, pp. 119-122.

[13] Walker, S. M., "Games of the Checkers Family in Line, Plane

and Space," *Bulletin of The American Mathematical Society*, Vol. 52, No. 9, September 1946, p. 825, art. 325.

[14] See Chapter V, Section 3a.

[15] The vertices B, D can be constructed as before. From the two congruent right triangles ABP, ADQ we have:

$$AB = AD, \qquad BAD = BAP + PAD = 90 \cdot$$

hence A, B, D are the three vertices of a square.

The fourth vertex is the symmetric C of A with respect to the mid-point of the diagonal BD.

Observe that if R is the projection of C upon the line APQ, we have $QR = AP$, for the two segments are the projections of two equal and parallel segments CD, AB upon the same line APQ.

Cf. Educational Times, Reprints, series 3, Vol. 5, 1918, p. 72, Q. 18570.

Chapter VI

MATHEMATICAL ASIDES

1 · MATHEMATICAL ASIDES

A · "It Is Obvious That . . ." It would be no exaggeration to say that in the writings on no other subject do the authors have as often recourse to the phrase, "it is obvious that," as is the case in mathematics. You have been frequently annoyed, no doubt, by this reference to the obviousness of certain statements which to you seem anything but obvious. Does that mean that the mathematical writer is so much smarter than his reader? Is the author indulging in a sadistic pleasure, or is he trying to poke fun at his reader? Perhaps a few examples picked more or less at random, might shed some light on this sore spot.

B · Four Examples A student missed one of the four quizzes given during the term. The instructor computed the student's average on the three quizzes taken. What grade did the instructor, by this procedure, grant the student for the quiz missed?

If a, b, c, are the grades made by the student, and x the grade the instructor granted for the quiz missed, we have

$$(a+b+c)/3 = (a+b+c+x)/4 = (a+b+c)/4 + x/4$$

hence
$$x = (a+b+c)/3$$

Thus the instructor granted the student, for the quiz missed, a grade equal to the average of the grades the student made on the quizzes he took.

After this result has been stated explicitly many a reader is likely to feel that the recourse to calculations was unnecessary, for this answer seem obvious. This is actually the case. All that is needed is to observe that since the "granted" grade does not alter the average, it must be equal to that average,

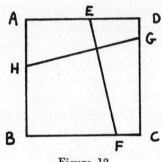

Figure 12

or may differ from that average by a couple of points, at most, in either direction, if fractions are to be taken into consideration.

As a second example consider the proposition: *If two perpendicular lines are drawn in the plane of a square, the segment intercepted by a pair of opposite sides of the square on one of the two lines is equal to the segment which the other pair of opposite sides intercepts on the other given line.*

Let *ABCD* be the given square and let the two pairs of opposite sides *AD, BC* and *AB, CD* intercept the segments *EF, HG*, respectively, on the two given lines (Fig. 12). We are to show that those two segments are equal.

The proposition may be proved in a number of ways of

varying degrees of complexity. But actually the proposition is practically obvious. Indeed, if we leave the line EF in its place and imagine that we spin the square, and the line GH with it, about the center O of the square counter-clockwise by an angle of $90°$, the sides DA, AB, BC, CD will occupy the present positions of the sides AB, BC, CD, DA, respectively, and the line HC which is perpendicular to EF, by assumption, will, on account of the rotation, become parallel to EF, and therefore $HG = EF$.

As a third example consider the problem: *A rigid ellipse moves so that it constantly remains tangent to the coordinate axes. Find the locus of the center of the ellipse.*

In general a problem of this type offers considerable difficulty and its solution may be long and laborious. In the present case, however, the solution is "obvious."

Indeed, the coordinate axes are a pair of rectangular tangents drawn to the ellipse, hence the origin O lies on the *Monge* circle (also called the *orthoptic circle* of the *director circle*) of the ellipse. Thus the distance of the origin from the center C of the ellipse is equal to the radius of the Monge circle, hence the locus of C is a circle equal to the Monge circle of the ellipse, and having O for center.

In order to come closer to the "obviousness" of an author we shall consider now one more example, the last, taken from this writer's own experience.

Given a tetrahedron (T), $ABCD$ and a point M, the four planes passing through the vertices A, B, C, D, and perpendicular to the lines AM, BM, CM, DM, respectively, form a tetrahedron called the *antipedal* tetrahedron (S) of (T) for the point M. Now let A', B', C', D', be the points in which the lines MA, MB, MC, MD, meet again the circumsphere (O) of (T), and let (S') be the antipedal tetrahedron, for the point M, of the tetrahedron (T') $A'B'C'D'$.

Considering poles and polar planes the writer arrived at the surprising conclusion that the lines joining the corresponding vertices of the two antipedal tetrahedrons (S), (S') meet in the center O of the sphere (O) and bisect each other. This seemed incredible, for the property does not involve the point M, while both tetrahedrons (S) and (S') depend on that point. The writer did not follow up this result.

Studying the same figure from an entirely new angle the writer stumbled again upon the very same result. It was not possible to doubt it any more. But if the result is valid, its simplicity would suggest that there must be a more direct approach to it than either of the two methods used hitherto. Further reflection brought the realization that the two faces of the two tetrahedrons (S), (S') which are perpendicular to the chord AMA' of the sphere (O) at the points A, A' are symmetrical with respect to the mediator (i.e., the perpendicular bisecting plane) of this chord. Now this mediator passes the center O of the sphere (O), hence the two planes are symmetrical with respect to the center O. The same holds for any other pair of corresponding faces of the two tetrahedrons (S), (S'). Thus the two tetrahedrons are symmetrical with respect to the center O, hence the proposition, which thus becomes "almost obvious" a priori.

$C \cdot$ *An Explanation*   An attentive scrutiny of the examples given reveals some very interesting features of what lies behind that so frequently troublesome statement "it is obvious that." To someone who is not familiar with the Monge circle of the ellipse the solution pointed out in connection with the problem of the ellipse is, of course, not obvious. The solver, before he arrives at an answer, will have to discover the Monge circle for himself, even if only in a round-about way, and in not a very explicit form. But even to a person to whom the Monge circle is not a novelty, the solution indicated may not occur very readily. The circle is not mentioned in the question. The pair of rectangular tangents constituted by the two coordinate axes is the only hint at that circle. For success in the solution of the problem this slight hint must be sufficient to evoke in the mind of the solver the image of the Monge circle and its relation to the problem. Whether that hint will suffice or not may depend upon a number of circumstances, like the degree of alertness of the solver at the moment, and other such conditions, but mainly upon the degree to which the Monge circle is fresh in the solver's mind.

This seems to point to the fact that there are what might be called degrees to which we may know a given fact. Should you ask a friend when Columbus discovered America he may not be able to supply the date. But he may nevertheless be able to recall the last two digits of that date if you would mention the first two digits, 1, 4. Should your friend not be able to do that either, and you would quote the full date 1492, he may agree with you very readily and assure you that he has known that date for many years. The situation seems to be akin to what the psychologists call the "threshold of sensitivity." The question is how much stimulus is necessary in order to make the solver aware of, to bring to the forefront of his mind, and to make available for use, a piece of information which is lying dormant somewhere in his memory, in his storeroom of knowledge. The fresher this information is the less stimulus will be necessary. Thus, if the solver dealt with the Monge circle very recently, he is much more likely to perceive the connection between that circle and the problem at hand than if he had no occasion to refer to it for a long time. If the circle is fresh in his mind and he is well grounded in its use, the solution is "obvious." Example one illustrates the same idea, perhaps, in a smaller way. To someone who deals with averages and statistical data the question may border on triviality, while to someone else the answer to the question becomes obvious only post factum.

When one has been working on a given subject any length of time one has the opportunity to see it from different angles and one becomes familiar with its various ramifications. The interconnections between the various parts become as natural, say, as the connection between two parts of the same familiar melody. One gets the feeling that these interrelations cannot possibly be missed or overlooked by anybody, so that it is quite superfluous to be pointing them out, and a statement like "it is obvious that" will be sufficient to put any reader on the right track. The good faith of the author using such an expression need not be questioned, but his optimism may have less secure foundations than he thinks. To the reader the connections between the parts of the subject may not become clear as rapidly as the author expects, and the author's "it is obvious that" may sound like mockery, or at least like an underserved reproach. The reader may, if he wants to, find consolation in the fact that, as time goes on, the subject may "cool off" in the mind of the author, the relations that were so vivid at the time of writing may fade away, and what was "obvious" may become to him incomprehensible and even impenetrable. It is no uncommon occurrence to witness scholars of renown discussing their own contributions and being "stumped" by the "obvious" in their own writings. While this may offer some solace to the harassed reader, it does not do away with the difficulty. The author is always confronted with the task of deciding what he should explain in detail, and what he can leave to the erudition of the reader. Unfortunately there can be no definite answer to this question. Superfluous verbosity may obscure the subject just as effectively as undue reticence.

The example with the antipedal tetrahedrons illustrates another aspect of the "obvious." It happens quite often that we obtain propositions by more or less laborious methods, only to discover that the result may be established by very simple reflections that are quite obvious, or nearly so. Why is it easier to arrive at these results by the devious route rather than by the simple one, the direct one? The reason is probably purely psychological. We just do not expect to get "something for nothing," and apply to our problems our standard tools in which we have confidence and which we know how to handle with some skill. If it happens that the simplicity of the result obtained does not seem to be in keeping with the heavy machinery we used to derive it, we may look around, discover an appropriately direct method, and end up by declaring *a posteriori* that the proposition was obvious *a priori*. More often than not this is done without any mention being made of the original method by which the proposition was actually derived in the first place. The example about the tetrahedron is "telling out of school."

D · Analogy as a Useful Guide to Discovery   The finding of new properties, of new propositions is a troublesome undertaking, whichever way you may look at it. There is no standard path at the end of which is the rainbow. This road of discovery, just as forbidding as it is alluring, is negotiated mainly by groping, clumsily and blunderingly. Of the meager sources of light available probably the best is provided by analogy. In the theory of functions of two or more variables we are guided by the ideas and results which proved successful in the study of functions of one variable, to mention a simple and familiar example. While studying three-dimensional geometry we keep an eye on the results available in the plane. This is particularly true when properties of the tetrahedron and the sphere are sought. It is an excellent indoor sport to extend to space known properties of the triangle and the circle. The efforts exerted in this direction are frequently rewarded very readily. Numerous examples of such extensions of properties of the plane may be found in the problem departments of the *American Mathematical Monthly* and the *Mathematics Magazine*. In many cases the generalization is so natural, so close at hand, that one wonders how the author of the property in the plane failed to think of the three-dimensional case. But let only those of us cast the stone of reproach who feel themselves without guilt. Have you thought of extending to space the property of the moving ellipse we considered a while back? The proposition applies nevertheless, and with only obvious modifications of the terms involved, to a rigid ellipsoid which moves so as to remain constantly tangent to the coordinate planes, and the locus of the center of the ellipse is a sphere having the origin for center and equal to the Monge sphere of the ellipsoid.

E · Limitations of That Method   Valuable as such analogies may be, one must not place too much reliance on them, for they are often misleading. Here is an illustration. Four coplanar, non-concyclic points, *A, B, C, D,* determine four circles, *ABC, BCD, CDA, DAB.* Now it is readily shown by the use of inversion, that if any two of these four circles are orthogonal, the remaining two circles are also orthogonal to each other. The extension to space is obvious. Five non-cospherical points *A, B, C, D, E,* determine five spheres *ABCD, BCDE, CDEA, DEAB, EABC.* If two of these five spheres are orthogonal, does it follow that the remaining three are mutually orthogonal? Perhaps this premise is too weak, and to be on the safe side it would be better to assume that three of the five spheres are mutually orthogonal and draw the conclusion that the remaining two are orthogonal to each other. Well, as a matter of fact the proposition is false in either case. More than that. Not only does the orthogonality of three of the spheres not imply the orthogonality of the remaining two, but, on the contrary, if three of the spheres are orthogonal, the remaining two *cannot be orthogonal* to each other.

The three projections, upon the sides of a triangle, of a point on the circumcircle of that triangle are collinear (the Simson line), and this is only true of the points on the circumcircle. It is easy to formulate the analogous proposition in space, but the analogy is distressingly misleading. There are points in space whose four projections upon the faces of the tetrahedron are coplanar, but these points have nothing to do with the circumsphere of the tetrahedron. The locus of those points is a cubic surface passing through the edges of the tetrahedron and through the quadritangent centers of the tetrahedron.

But one does not have to go so far afield to find striking examples when the analogy between the plane and space fails to hold in places where one would confidently expect it to be the case. That the cube in space is the analogue of the square in the plane requires no argument. Let us follow up this analogy by considering the diagonals of the two figures. The diagonals of a square are equal. So are the diagonals of a cube. The diagonals of a square bisect each other. The same holds for the diagonals of a cube. The diagonals of a square are perpendicular. Is that true of the diagonals of a cube? No, it is not. The diagonals of a cube cannot be mutually perpendicular, for if they were we would have at their common point four mutually perpendicular lines which, in a three-dimensional space, would be quite a spectacle to behold.

The breaking down of the analogy between the plane and space is not the only shortcoming of this source of inspiration in our quest for new propositions in space. Three-dimensional space is much richer in relations than is the two-dimensional plane. It has properties which have no counterpart in the plane. The analog of curves in the plane are surfaces in space. But in addition to the surfaces we also study curves in space, both plane and skew. If we would rely too closely upon the plane, specific properties of space might be overlooked. We shall now consider an example in which both the help and the limitations of the plane consideration come into play.

Our school books on geometry told us that a triangle has three altitudes and that those three lines have a point in common. Contrary to what is implicitly believed, this proposition is not to be found in Euclid's Elements. But it would be erroneous to conclude from this absence that Euclid did not know this property. We are certain that Archimedes who lived soon after Euclid, was aware of this property of the triangle. He refers to it in his writings as to something his readers are expected to be familiar with.

F · The Altitudes of a Triangle and of a Tetrahedron A triangle is a plane figure formed by three non-concurrent lines. Its analog in three dimensional space is a figure determined by four planes having no point in common, that is, a tetrahedron, or what is the same thing, a triangular pyramid. This solid was familiar to the Greeks from the earliest times. They learned of it from the Egyptians who were building pyramids many centuries before the Greeks appeared on the scene of history. The Greeks considered the altitudes of the tetrahedron in connection with the formula for the volume of that solid.

A tetrahedron has four altitudes. The analogy with the triangle suggests the obvious question: "Do the four altitudes of a tetrahedron meet in a point?" It would take a hardy soul, indeed, to maintain that this simple idea never occurred to the inquisitive minds of the Greek geometers. On the other hand, in all the mathematical writings of the Greeks which came down to us in any form, there is not the slightest hint that the question of the concurrence of the altitudes of a tetrahedron has ever preoccupied those imaginative scholars. Moreover, this question was ignored, with a persistence worthy of a better cause, by the middle ages, the Renaissance, and clear down to the nineteenth century.

It is, however, possible to find some attenuating circumstances for this curious silence. If one is asked to answer our question by a "Yes" or "No" one is faced with the same uncomfortable situation as when confronted with the question: "Will you stop beating your wife?" One would be in the wrong no matter which of the two alternatives one decides to espouse.

The secret of this puzzling situation is simple. There are types of tetrahedrons whose altitudes meet in a point, and there are types for which this is not the case. If the altitudes of a tetrahedron meet in a point, that entails the property that each edge of the tetrahedron makes a right angle with the opposite edge. Conversely, if a tetrahedron has the latter property, it also has the former. This type of a tetrahedron, commonly referred to as "orthocentric," has a good many other properties which are close analogs of properties of the triangle.

Three mutually orthogonal planes, like, for instance, the floor and two adjacent walls in a room, and any fourth plane form a "trirectangular" tetrahedron. The line of intersection of any two or the first three planes considered is perpendicular to the third plane and is therefore an altitude of the tetrahedron. Thus the four altitudes of a trirectangular tetrahedron all pass through the vertex of that tetrahedron common to the three mutually orthogonal faces of that solid.

A tetrahedron ABCD may have only one pair of mutually orthogonal opposite edges, say, AB and CD. In such a "semi-orthocentric" tetrahedron the altitude issued from A and B have a point in common, and the same holds for the altitudes issued from C and D.

The three types of tetrahedrons considered above is all that we can get out of the analogy between the altitudes of the triangle and the tetrahedon. This, however, does not exhaust the topic, for there are tetrahedrons in which no edge is perpendicular to the opposite edge, and therefore none of the four alitudes meets another altitude, that is, the four altitudes are four mutually skew lines. Does that mean that in such a case the four altitudes are four totally unrelated lines? No, the four altitudes are not total strangers to each other, but their relation to one another is not an analogy of a property of the altitudes of a triangle.

In order to show what that relation is we have to consider some preliminaries. A given point M and two skew lines a, b, in space determine two planes $(P) = M{-}a$, $(Q) = M{-}b$ which have a line u in common (Fig. 13). The point M lies on their common line u. Furthermore, the two lines u and a lie in the plane (P), hence they have a point in common (we neglect the special case of parallelism), and the same holds for the lines u and b, for similar reasons. The upshot of the story is that we have constructed a line u passing through the point M and intersecting the lines a and b.

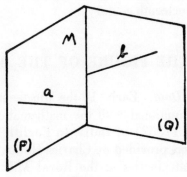

Figure 13

Suppose now that through the point M we draw a line c skew to the lines a and b. If we should treat other points of the line c the same way as we proceeded with the point M we may construct an infinite number of lines meeting the three skew lines a, b, c.

Let d be any line in space skew to each of the lines a, b, c. How many of the infinite number of lines meeting the latter three lines meet also the line d? It is proved that the answer to this query is: not more than two lines, as a rule. And the latter qualifier "as a rule" is the crux of the matter. There may be exceptions to the rule. And if it should happen that of that infinity of lines *three* lines should meet the line d, all the rest of them will do likewise. Thus the four mutually skew lines *a, b, c, d* will be met by an infinite number of straight lines. This is usually stated more succintly by saying that the lines *a, b, c, d* form a "hyperbolic group."

Let us now return to our four altitudes. We are now ready to state the proposition: "The four altitudes of a tetrahedron are four skew lines such that a line which meets any three of them also meets the fourth." The proposition may be stated more briefly: "The four altitudes of a tetrahedron form a hyperbolic group."

This proposition was first formulated during the third decade of the nineteenth century.

2 · "THE FIGURE OF THE BRIDE"

A · Historical Data   Early in the nineteenth century the West became acquainted with the mathematical writings of Bhaskara (1114-1185?) of India. In English a first glimpse of these works was provided by Charles Hutton (1737-1823), Professor of Mathematics at the Royal Military Academy, Woolwich. In 1812 Hutton published in London a three volume collection of "Tracts on Mathematical and Philosophical Subjects." Tract No. 33, vol. II, pp. 143-305 deals with "The History of Algebra of all Nations." In particular, pages 151-179 are devoted to "Indian Algebra." The revelations about Indian mathematics made therein must have created quite a stir at that time, considering that shortly after Hutton's work appeared, a professor of mathematics at the Royal Schools of Artillery on the other side of the English channel, namely O. Terquem (1782-1862), translated the part of Hutton's "History" relating to India and published it in Hachette's "Correspondance sur l'Ecole Polytechnique."[1]

B · The Theorem of Pythagoras in India   One of the works of Bhaskara which Hutton quotes and comments upon extensively is entitled "Lilavati" and is devoted largely to Arithmetic. In it there is the passage: "In the margin of the original, as here annexed, is drawn a figure of four equal right triangles joined in the manner indicated (Fig. 14) exhibiting a new and obvious proof of the 47th proposition of Euclid I (that is, the Pythagorean theorem): for here are the four right triangles,

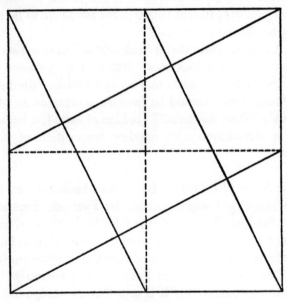

Figure 14

which are equal to twice the rectangle of their two perpendicular sides, and which together with the small square in the middle, being the square of the difference of those two sides, make up the large square on the hypotenuse (In modern notation: if a, b are the perpendicular sides of one of the triangles, the area of the big square is equal to $4 \cdot ab/2 + (a-b)^2 = a^2 + b^2$). Therefore the square on the hypotenuse is equal to the sum of the squares on the other two sides."

"And this may be considered the Indian demonstration

of the celebrated property of the sides of a right angled triangle; a property so much employed by their geometricians,
and so often referred to in their writings by the name of 'the
figure of the bride' and 'the figure of the bride's chair' and
'the figure of the wedding chair', epithets which we may conjecture have been suggested by the above figure bearing some
resemblance to a palanquin or a sedan chair, in which it is
the usual practice, in that country, for the bride to be carried
home to her husband's house."

Is it quite certain that a "celebrated" proposition necessarily has to have a nickname? But if it be assumed that it
should, then it would seem that an epithet like "the figure of
the wedding chair" should be fully as acceptable as, say, the
well known "Pons Asinorum" (bridge of asses), a name often
quoted in connection with another proposition of Euclid's
Elements.

C · *The Story of Lilavati*   Bhaskara's book *Lilavati* (meaning: the beautiful) was translated by Fyzi into Persian, "by
order of the king." In his preface to the book, Fyzi narrates
a story connected with the origin of the book. Hutton finds this
account to be "very curious, and containing some useful par-
ticulars" and therefore he includes it "as a postscript" at the
end of his own narrative. "It is said that the composing of the
Lilavati was occasioned by the following circumstance. Lilavati was the name of the author's (Bhaskara's) daughter, concerning whom it appeared, from the qualities of the Ascendant
at her birth, that she was destined to pass her life unmarried,
and to remain without children. The father ascertained a lucky
hour for contracting her in marriage, that she might be firmly
connected and have children. It is said that when that hour
approached, he brought his daughter and his intended son
near him. He left the hour cup on the vessel of water, and
kept in attendance a time-knowing astrologer, in order that
when the cup should subside in the water, those two precious

jewels should be united. But, as the intended arrangement was not according to destiny, it happened that the girl, from a curiosity natural to children, looked into the cup, to observe the water coming in at the hole; when by chance a pearl separated from her bridal dress, fell into the cup, and, rolling down to the hole, stopped the influx of the water. So the astrologer waited in expectation of the promised hour. When the operation of the cup had thus been delayed beyond all moderate time, the father was in consternation, and examining, he found that a small pearl had stopped the course of the water, and that the long-expected hour was passed. In short, the father, thus disappointed, said to his unfortunate daughter, I will write a book of your name, which shall remain to the latest times—for a good name is a second life, and the groundwork of eternal existence."

Bhaskara thus wrote his book *Lilavati* in fulfillment of a promise given his beautiful daughter when he found out by the stars that she was fated to spinsterhood. The account may may be history, or it may be legend. Most likely, it is a mixture of both truth and fancy, in unknown percentages. What is certain, however, is that the work Bhaskara produced under the title *Lilavati* will "remain to the latest times," as a document in the history of culture.

3 · RUNNING AROUND IN CIRCLES

When watching the popular game of "Pinning the tail on the donkey" we are often amused, and not a little surprised to see the blindfolded performers instead of making straight for the object sought, wander off to one side or the other. However, these defenseless victims of our derision do no worse than they could be expected to. In fact they would go much farther astray, if the "donkey" were placed at a greater distance from the point where the chase begins.

The beautiful San Marco cathedral in Venice is about
ninety yards wide, and the square in the front of it (Piazza
San Marco) is nearly two hundred yards long. Those who
attempt to reach the cathedral, blindfolded, starting from the
end of the square directly opposite the building, find themselves at either the right side or the left side of the square.
None of them ever reaches the cathedral.

The Russian mathematician, Y. I. Perelman, tells of one
hundred aviation cadets who were lined up, blindfolded, at
the edge of an airfield and ordered to walk straight ahead.
The young men started out as they were bid, but they could
not keep it up. After a short while they began to turn to the
side, some to the right, some to the left. They actually walked
in circles, each of them repeatedly crossing his own tracks.

Mother Nature numbers in her vast arsenal of tricks
quite a few blindfolding devices: pitch dark nights, dense
fogs, blinding snowstorms, thick forests, trackless open spaces
—deserts, large bodies of water, etc. When a hapless traveler,
trapped by the merciless elements, is deprived by them of his
sense of vision, he is unable to follow any fixed direction, and
"runs around in circles."

Fright may have an equally disastrous effect upon a
man's ability to orient himself. When fleeing from his pur-
suers, a tracked man believing to be running straight ahead
and away from danger, actually runs in circles.

These facts have been known for a long time. They have
frequently been exploited by writers of fiction. Leo Tolstoy,
thoroughly familiar with the snowstorms of the vast Russian
plains, as well as with the folklore connected with them, has
more than once described the aimless wanderings of people
lost in the snowy deserts.

Stories of this kind form a part of the lore of the American cowboy.[2]

The late American playwright, Eugene O'Neill, in his
powerful drama *Emperor Jones* describes the flight of the

horror stricken "emperor" through a forest, at night. The renowned author bases the climax of his play on the fact that after a night of frantic running the maddened fugitive is overtaken at the spot where he entered the forest the evening before. One of the characters of the play remarks, knowingly, in good cockney dialect: "If 'e lost 'is way in these stinkin' woods, 'e'd likely turn in a circle without 'is knowing it. They all does."

The tendency to move in a circle or circles, when the controlling action of the eye is inoperative, is not an exclusive characteristic of man. Animals behave likewise. When a chicken loses its head, literally, it runs around in circles, as the proverbial saying has it. A blindfolded dog swims in circles. Blind birds fly in circles.

Hunted animals when consistently pursued, end up by running in circles. As reliable an observer as Roy Chapman Andrews, of the New York Museum of Natural History, in his article, "The Lure of the Mongolian Plains,"[3] testifies to "the fatal desire (of the antelope) to turn in a circle about the pursuer."

Toward the end of the last century the Norwegian biologist, F. O. Guldberg devoted considerable attention to the question of circular motion in man and animals. He collected a good deal of authenticated material bearing upon the subject.

He tells of three travelers who during a snowy night left the shelter of a woodman's hut in an attempt to reach their home, located on the opposite side of a valley, about three miles wide. They started out in the proper direction, but after a while they deviated from it, without realizing the change. By the time they estimated that they should have reached their destinaiton, they discovered that they were once again close to the very hut they so imprudently had abandoned. Undaunted by this disappointment they started out again— with the same unfortunate result. The third and the fourth attempts both had the same disheartening outcome. They came back to the very same hut, as though under some magic spell, as though tied to it by an invisible chain. When even the fifth try brought no better luck, our tired travelers arrived at the conclusion that it might be the better part of valor to wait for the light of day.

Guldberg has similar well substantiated stories about rowers in the open sea who try to reach a point on the shore during a dark starless night or during a fog. Thus rowers who undertook to cross a sound three miles wide during foggy weather, never succeeded in reaching their goal. Without knowing it they described two circles. When they finally came ashore they discovered to their great amazement that it was the spot they started from.

On the strength of such information Guldberg contributed an article to a biological magazine in which he discussed the topic: "Circular Motion as the Basic Motion of Animals."[4]

When daddy crouches down on the floor in order to wind up junior's mechanical automobile, for the amusement of the boy, and no less his own, the entertaining and perverse plaything seldom chooses to follow the straight and narrow path lying directly ahead, but instead describes some kind of arc, away from the line of virtue. These extravagances of junior's toy may seem strange and capricious, if not vicious. But a little reflection will readily explain the puzzle of the little vehicle's behavior.

In order that the propelled toy run along a straight line, it is necessary that the wheels on the two sides of it shall be strictly of equal size. If they are not, the little automobile will turn to the side of the smaller wheels. There is no reason for suspecting the plaything of willful misconduct. But does not its behavior, whether wicked or not, offer a clue to the mystifying stories of human misadventures which we have described?

Under ordinary circumstances a man, while walking, "watches his step," and "looks where he is going." He needs the help of his senses, principally his eyes, to get to the point he intends to reach. But when these controls are not available, the pedestrian will follow the direction in which he started out, only if the length of the step he takes with one foot is exactly equal to the length of the step he takes with the other. Is this equality of the steps a thing that may be taken for granted? In the vast majority of people the muscular development of the two legs is not the same, it is therefore to be expected that the steps will be uneven, rather than the contrary. To be sure, we are not aware of this difference, for the good reason that it usually amounts to very little. But small as it may be, it brings about some very striking consequences.

If the right and left steps were strictly equal, the tracks of the two feet would lie on two parallel lines, a certain distance, say w, apart. But suppose there is a difference, say d, between the length of the right and the left step; let us assume that the difference is very small, say d amounts to no more than 1/200 of an inch. After twenty thousand steps with each foot the difference of the distances traveled by the two feet will amount to 100 inches, which is nearly three yards.

Now, if the two feet are moving along two parallel lines, such an outcome is patently absurd: one foot cannot remain three yards behind the other. The difficulty vanishes on the assumption that the two feet move on two concentric circles.

The difference between the radii of the two concentric circles is the distance, w, between the tracks of the two feet. Thus if the smaller circle has a radius R, the larger circle has a radius $R+w$. The lengths of the two circumferences of the two circles are, respectively $2\pi R$ and $2\pi(R+w)$ according to a well known formula. The difference between the total distances traveled by the two feet while describing the two circles is thus $2\pi w$.

If a pedestrian moves in a circle having a radius equal to one mile (as was approximately the case with the three travelers who tried to cross the valley), how much difference is there in the steps of his feet? The length of the circumference of the circle is $2\pi.12.5280$ inches. If we take the length s of one step to be 27 inches, the pedestrian made in all $2\pi.12.5280/27$ steps. With each foot he made $2\pi.12.5280/2.27$ steps. If in the expression $2\pi w$ we take $w = 4$ in., we come to the conclusion that in $2\pi.12.5280/2.27$ steps one foot covered a distance of $2\pi.4$ inches longer than the other. If we divide the latter number by the former we obtain the difference between the lengths of the steps of the two feet of the pedestrian. The actual computation yields the surprising result of less than 0.01 part of a third of an inch. And this triflng difference was enough to keep our intrepid and unlucky travelers out of their home!

An analogous argument may enable us to establish a relation between the difference, d, of the steps and the length, R, of the radius of the circle which the pedestrian will describe. The length of the circumference of radius R is $2\pi R$. If s is the length of a step, the pedestrian will make $2\pi R/s$ steps all told. With one foot he will make $2\pi R/2s$ steps. If d is the difference between the steps, the foot making the longer step will cover an additional distance of $2\pi Rd/2s$. This additional distance is equal to $2\pi w$, as we have seen before. We have thus the equation

$$2\pi Rd/2s = 2\pi w$$

or

$$Rd = 2sw.$$

If we put $s = 27$ inches and $w = 4$ inches, we have

$$Rd = 216,$$

where both R and d are to be given in inches. This formula sohws that R and d are inversely proportional. Furthermore, it enables us to compute either R or d, if the other is given.

The development of the muscles in a man's two arms is not any more the same than that of his two legs, hence his strokes, when he is rowing, are of unequal efficacy, and his boat will move in a circle when he is unable to control his course with the help of his sight.

Similarly for the strength of the wings of a bird, and so on. This takes the mystery out of our story.

But we have not come to the end of that story. About a quarter of a century ago an American biologist, the late Asa A. Schaeffer, at the time professor of zoology at the University of Kansas, imparted a new twist to our problem.

Guldberg had already noticed that the repeated circles of our travelers in the snowy valley and of other analogous cases fall into a pattern which looks like a clock-spring spiral. Now Schaeffer believed that lower organisms, like amebas, which move in three dimensions, are governed in their motion by a mechanism which makes them travel along a helical spiral. This mechanism, he believed, survives in higher animals, including man. Hence blindfolded persons walk, run, swim, row, and drive automobiles in clock-spring spiral paths, of greater or less regularity, when attempting a straightaway.

Schaeffer conducted a large number of experiments involving many people. He recorded his results carefully, with all the refinements of modern experimental technique. He published his findings in a paper more than a hundred pages long, in the *Journal of Morphology and Physiology*,[5] under the title: "Spiral Motion in Man."

Schaeffer rejected the older theory and its "simian simplicity." But even if the correctness of Schaeffer's own theory be granted, it does not seem that this necessarily invalidates Guldberg's explanation. The two causes may complement each other and may be operating simultaneously.

4 · TOO MANY?

When Casper entered the den of his master on that bright and cheerful morning, he was a bit surprised to notice an unfamiliar object on the desk—an elegant box of small size. Mr. Purefoy must have brought it when he came home late last night; or was it a surprise gift from Mrs. P.? Or . . . but wherever it came from Casper would be amiss in his duties as a butler, if he did not examine the unfamiliar object very carefully. And then open it. The latter operation turned out to be less difficult than could be anticipated. The box contained a new brand of cigarettes, packed very carefully.

Of course, the butler's first impulse was to have a puff at one of those newfangled things. But, on second thought, Casper hesitated. The top of the box consisted of a layer of twenty cigarettes neatly and tightly placed one alongside the other. The absence of one of them would be all too conspicuous. After earnest deliberation Casper arrived at the conclusion that the tasting of new brands of cigarettes did not fall within the scope of his duties. He left the cigarettes where they were, closed the box, and returned to the examination of the outlandish figures on the little box.

Casper's deep absorption in his esthetic contemplations was brought to an abrupt end when suddenly the lid of the box sprang open, and most of the cigarettes found themselves on top of the desk, and some underneath. What the butler muttered under his breath at this occurrence may or may not have been fit to print, but whatever it was it had no appreciable effect upon the situation. And something had to be done!

Cross and disgruntled, the butler emptied the remaining contents of the box upon the table, picked the cigarettes off the floor, and went to work. He quickly put twenty cigarettes in a row, one beside the other, on the bottom of the box. This covered the bottom completely. On the top of this layer he placed another layer of twenty cigarettes. Casper worked diligently. He was careful to keep the layers smooth and even. When he completed the eighth layer, he breathed a sigh of relief; he was pleased with himself and his handiwork. The box was full and looked exactly as when he first opened it.

"You could not see any difference to save your life," he flattered himself.

Casper lowered the lid upon the box and was ready to put it in an appropriate place, when he noticed on the desk, behind the box, several cigarettes. He blinked, "Am I seeing double?" he asked himself, bewildered. But the cigarettes were real. There was no "maybe" about it. He counted them. Sixteen cigarettes, of the same brand as those in the box. He counted again, he counted them once more—the same sixteen, no more and no less. He half-heartedly bent down to look again under the desk, lay down on the floor to peep under the other pieces of furniture—all in vain; no other cigarettes. Just sixteen. Not enough for another layer. "But if I had a full complement of twenty," he argued with himself, "that would do no good, either. There just is no room in the box for another row." Casper had as full a box as he should have, and sixteen cigarettes on top of that. It should not have been so, but it was. Casper may never before have experienced *embarras des richesses*, but he did that morning and it was not to his liking.

In the evening of the same day, after dinner, when Mr. Purefoy retired to his den, there was a knock on the door.

"Anything very urgent?" asked the boss impatiently, when the butler stepped into the room.

"Those extra cigarettes, Sir. They are in the side drawer of your desk," Casper reported in a subdued and unusually meek voice.

Mr. P. looked at his butler in astonishment. "What are you talking about, Casper? I brought home a box of cigarettes, this box in fact." He pointed to the box so familiar to Casper. It was standing there with the lid raised, but otherwise undisturbed. "But I brought no extras of any kind, that I know."

"That is quite correct, Mr. Purefoy. But when I was arranging your den this morning I somehow upset this box, and the cigarettes fell out. I repacked the box and filled it to capacity; there were sixteen cigarettes left for which there was no room in the box."

The boss broke out in a hearty laugh. "You should have tried the trick again, Casper", he said. "If with each repacking you could save sixteen cigarettes, you . . . ," and he started laughing again.

"I repacked that box three times in a row, Sir, but no such thing happened again. I got in eight rows of twenty cigarettes each time, and no more, no less. The thing has had me worried all day long. I am not a superstitious man, but these extra sixteen cigarettes give me the creeps, Sir."

By this time it dawned upon Mr. Purefoy that to the man in front of him the accident with the cigarettes was no laughing matter.

"Well, Casper," he said reassuringly, "right now, as you see, I have before me some urgent papers that I must go over. But tomorrow morning, if you examine this box carefully, you may perhaps find an answer to the question that has been bothering you." "But be sure," he admonished the butler when he was on the way out, "to leave the cigarettes in the box in the same order you find them there."

Casper's sleep that night was quite disturbed. He dreamed of boxes, large and small, carried by raging flood waters, of burning stacks of white logs bellowing with dense smoke. . . He was himself trying to rescue those logs by pushing them into the boxes carried by the flood. . . .

Next morning he grabbed the very first opportunity to get close to that confounded box. When he raised its lid, there was the row of twenty cigarettes, as though nothing ever happened. He removed that layer, being very careful not to disturb the cigarettes underneath.

When Casper put the twenty cigarettes aside and took another look at the box, the scenery was entirely different. The next row had only nineteen cigarettes, and they were placed in the grooves formed by the adjacent cigarettes of the layer below, which layer consisted of twenty cigarettes. "I'll be darned," said Casper aloud, and as though in spite of himself. He continued to remove layer after layer. By the time he reached the bottom of the box he had counted up four layers of nineteen cigarettes each, sandwiched in between five layers of twenty cigarettes each.

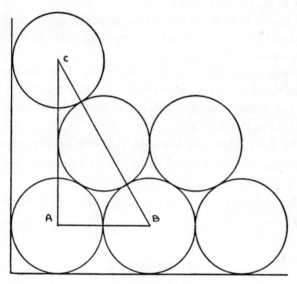

Figure 15

"For once I am forced to admit that the boss is right," Casper pensively murmured to himself. "They are all here, down to the very last of the one hundred and seventy six of them."

That evening, when Mr. P. happened to come upon his butler, he asked him, with a perceptible touch of mockery in his voice, "And those extra cigarettes, Casper, what about them?"

"There is nothing extra about them any longer, Sir. They are all alike now, snug in the same box."

"You may perhaps be interested to know," was the parting dart Mr. P. threw over his shoulder at his butler, "that the box with all the cigarettes in, is not quite as full as when you packed it your way, leaving out the 'extras.' "

Explanation. If the front wall of the box were transparent, the cigarettes would appear to us as little circles, each tangent to all the adjacent circles.

If the radius of such a circle is a, the distance AB between the centers A, B of the first two cigarettes in the lowest layer is equal to 2a. (Fig. 15)

The distance BC from B to the center C of the first cigarette in the third layer up is equal to $a+2a+a=4a$. Hence, (from the right triangle ABC) the vertical distance $CA = \sqrt{(4a)^2-(2a)^2}=2a\sqrt{3}$.

The four distances between the five twenty cigarettes rows are thus together equal to $8a\sqrt{3}$, and the total height of the stack is $a+a+8a\sqrt{3}=15.9a$.

The eight layers of twenty cigarettes each form a stack 16a high.

FOOTNOTES

[1] Vol. III, No. 3, January 1816, pp. 259-283.

[2] See, for instance, Holling, Holling C., *The Book of Cowboys* (New York, 1936), Chapter 32.

[3] "The Lure of the Mongolian Plains" Roy Chapman Andrews, *Harper's Magazine*, Vol. 141, 1920.

[4] "Circular Motion as the Basic Motion of Animals," *Zeitschrift fuer Biologie*, Vol. 35, 1897, pp. 419-458.

[5] Vol. 45, 1928, pp. 293-298.

Chapter VII

MATHEMATICS AS RECREATION

1 · MATHEMATICAL FOLKLORE

Introduction Teachers may flatter themselves that the task of teaching school is their exclusive privilege. But this is far from being the case. A considerable amount of teaching is done by the pupils. For better or worse, we learn a good deal from our schoolmates. Occasionally we learn from them even some mathematics, a kind of mathematics for which the teachers have no time and no patience, not to say no use. This kind of mathematics usually consists of puzzles or riddles, which are very simple in their statement. Their solution calls for almost no learning, no erudition, but requires of the solver a considerable effort of imagination and quite a bit of ingenuity.

The pupil who brings such a riddle to class usually learned it himself by word of mouth from someone who in turn learned it the same way, so that it may quite appropriately be said to be mathematical folklore. Other reasons for the use of this appellation may be gleaned from the text that is to follow.

A · "River Crossing" Problems   The most striking of those riddles, the one that appealed to me most those many years back, has to do with the wolf, the goat and the cabbage. The story runs something like this. A boatman undertakes to ferry a wolf, a goat, and a basket of cabbages across a river. His boat is so small that there is room for himself and either the wolf, or the goat, or the basket of cabbages, but no more. How is he to accomplish his task without loss or damage to the property that was intrusted to him?

In this riddle the most troublesome passenger, from the boatman's point of view, the one that "gets his goat", is, of course, the goat. If the man should start by taking across the basket of cabbages, there may not be much left of the goat when he comes back. Should he take the wolf first, he is likely to find upon his return that the goat has done considerable damage to the supply of cabbages.

One may be inclined to inquire who first imagined this very original puzzle. It is just as natural to ask this question as it is difficult to answer it. What is certain is that the riddle is hoary with age. It was known in the Orient long before the Christian era. In the West it may be traced as far back as the eighth century, to a book written by Alcuin or Flaccus Albinus (c. 735-804,), an English educator and ecclesiastic who lived at the court of Charlemagne and was in charge of education in this ruler's vast Frankish empire.

The simplest, in fact the only way out, is to take along the goat first. But what next? Whether he takes next the wolf or the basket of cabbages he will be in exactly the same predicament as he was before, by the time he attempts to return for the third item of his load. It is here that the boatman had a brilliant idea. When he brings the wolf across as item number two, he takes the troublesome goat back with him to the first shore, leaves him there all by himself and ferries the cabbages across; then he comes back for the goat, and the job is done.

Those who busied themselves with this ancient puzzle deemed the boatman's idea of ferrying the goat forth and back so striking that they paid him the highest possible compliment: they tried to imitate him. They tried to make up puzzles the solution of which involved the same idea. Here are some examples.

A group of soldiers wish to cross a river. They spy a boat with two boys in it. Either boy can operate the boat. But the boat is so small that it can carry at most one soldier or the two boys. The soldiers got across. How did they manage it?

In the Smith family father and mother weigh in the neighborhood of 160 lbs. each, while John and Mary tip the scale at half that weight. On an excursion the whole family and their spaniel dog, Tom, weighing about a dozen pounds, have to cross a river in a boat about which they were warned that when loaded beyond 160 lbs. it becomes definitely unsafe. John, who was a bright boy and, besides, knew the story about the wolf, the goat, and the cabbages, found a way out of the difficult situation. Of course, it occurred to no member of the family to throw Tom into the water and let him get across under his own steam: the poor thing might catch a cold in the process.

Two jealous husbands and their wives must cross a river in a boat that holds only two persons. How can it be done so that a wife is never left with the other woman's husband unless her own husband is present?

The ambitious reader is not likely to have any more trouble with this problem than with the last two mentioned before. He will find that the crossing can be accomplished in five steps.

The problem becomes much more complicated when there are three couples, and the third husband is just as jealous as each of the first two. The task may be accomplished in the following manner.

Two of the three women go across, one returns, and takes

across the third one. When one of them returns, she remains with her husband, while the other two men go across to their wives. Next one of the two couples returns, the wife remains, and the two men go across. The only woman that is there goes across to bring with her one of the wives, and then goes back again to bring the third woman, but it would be more chivalrous for the husband of that third woman to go across to bring his wife over.

The successive steps may be arranged in the following table, where A, B, C represent the husbands, and X, Y, Z their respective wives.

	First bank	Second bank
	A, B, C; X, Y, Z	nobody
1°	A, B, C; X	Y, Z
2°	A, B, C; X, Y	Z
3°	A, B, C;	X, Y, Z
4°	A, B, C; X,	Y, Z
5°	A, ; X,	B, C; Y, Z
6°	A, B, ; X, Y	C; Z
7°	X, Y,	A, B, C; Z
8°	X, Y, Z	A, B, C
9°	Z	A, B, C; X, Y
10°	X, Z	A, B, C; Y
11°	nobody	A, B, C; X, Y, Z

Being in possession of the solution for three couples, it might be suggested to the reader to try it with four couples. Such a challenge, however, would be nothing less than a sadistic pleasure. For if you undertake the job and find a solution, your solution will be wrong. However, this would not be much of a humiliation, for you would be in good company. A renowned Italian mathematician of the 16th century, N. Tartaglia (1500-1557), also found a solution, and the solution was wrong. Yes, great mathematicians also make mistakes.[1] The secret in the matter is that the problem with four couples has no solution. Maurice Kraitchik, in his book, *Mathematical Recreations* (New York, 1942) has considered the problem for any number of couples. He shows that with a boat accommodating three persons the problem can be solved for five couples, but not for six or more.

Before we quit this topic it may not be out of place to ask: is this mathematics? If your answer is Yes, then it is contrary to the common conception that mathematics consists in figuring, in long and involved computation. If your answer is No, then how is one to account for the fact that those questions attracted and intrigued mathematicians primarily, even eminent ones among them; and the solutions of the problems were furnished by mathematicians?[2]

B · Multiplication Performed on the Fingers Among the most prized pieces of information gathered by the writer via the folklore route is the secret of a mechanical multiplication table, or to be more precise, of the more advanced, the more difficult part of that table. The secret is the more surprising in that one always has with him the necessary tools to make use of that mechanical table. Indeed, all the requisite machinery consists of a complete, unabridged set of one's fingers, thumbs included.[3] The preliminary mastery of the multiplication table of numbers not exceeding five would be of help.

We assign the number 6 to the little finger, on each hand, 7—to the ring finger, 8—to the middle finger, 9—to the pointer, and 10—to the thumb. We are now set up in business.

If you want to multiply, say, seven by nine, put your two hands before you, palms in, and put the tip of the ring finger, value 7, of one hand, say, the left hand, against the tip of the pointer, value 9, of the other hand. The two fingers thus joined and those below them are six in number and they count for sixty towards the final result. Above the two joined fingers remain three fingers on the left hand and one finger on the right hand. Multiply those two numbers and add the product three to the value sixty we have already, and you have the required result. Thus: $7 \times 9 = 6 \times 10 + 3 \times 1 = 63$.

Let us do it once more, to make sure. To multiply, say, 6 by 8 put the little finger of the left hand against the middle finger of the right hand. The two joined fingers and those below them are four fingers, and they count for forty towards the final result. Above the two joined fingers there are four fingers on the left hand and two on the right hand; multiply those two numbers. The final result is: $6 \times 8 = 4 \times 10 + 4 \times 2 = 48$. With a little practice it is possible to read the result almost instantly.

Besides its arithmetical uses, this clever trick may also serve, with telling effect, to enhance the prestige of an ambitious grandfather in the eyes of a bright fourth-grade grandson. Strange to say, this secondary virtue of the ingenious arithmetical device completely escaped my notice when I first became acquainted, folklorewise, with the primary purpose of the artifice.

This remarkable scheme is a relic of remote antiquity. It is a part of a very elaborate method of digital computation developed in the Orient probably before the invention of writing and extensively used in classical antiquity. The method is frequently alluded to in the writings of the latter period On the other hand, this particular method of multiplication has survived until the present day. Competent observers report that it is still resorted to by the Wallachian peasants of southern Rumania.[4]

This tricky method of multiplication is, of course, a purely empirical discovery. The mathematical basis of its puzzling success is the fact that the equality:

(p) $(5+x) \; (5+y) = 10(x+y) + (5-x) \; (5-y)$

is an identity.

This identity may also be put in the form:

(q) $(5+x) (5+y) = 5(x+y) + xy + 5^2$,

which may perhaps be simpler but does not exhibit as clearly its digital origin.

If in (p) we replace 5 by a , where a is any number, we obtain the identity:

(r) $(a+x) (a+y) = 2a(x+y) + (a-x) (a-y)$,

which may also be written in the form:

(s) $(a+x) (a+y) = a(x+y) + xy + a^2$.

If in the identity (r) we replace a by 10, we obtain the formula:

(t) $(10+x) (10+y) = 20(x+y) + (10-x) (10-y)$,

which may also be written as:

(u) $(10+x) (10+y) = 10(x+y) + xy + 10^2$.

If we interpret (t) in a manner analogous to the interpretation we have for (p), we may use (t) for the multiplication of numbers within the range from 11 to 15.

The process may be continued by replacing a in (r) successively by 15, 20, 25, . . . and using the resulting identities in a manner analogous to the way we used (p) and (t).

That the masters of digital computation ever used these generalizations is quite unlikely.

C · "Pouring" Problems · The "Robot" Method Another type of problem which I recall having learned from my schoolmates is the following. The contents of a cask filled with 8 quarts of wine is to be divided into two equal parts using only the cask and two empty jugs with capacities of 5 quarts and 3 quarts respectively.[5]

This, too, is a riddle many, many centuries old, exactly how old is difficult to say. A solution may be arrived at by trial and error. The number of attempts necessary will be considerably reduced if a record is kept of the trials attempted in the form of, say, 3, 5, 0, which would mean that from the 8 quart cask we filled the 5 quart jug, and so on.

First solution

Vessel	Amount of wine in each vessel, by stages								
Stages	1	2	3	4	5	6	7	8	9
8 quart cask	8	5	5	2	2	7	7	4	4
5 quart jug	0	0	3	3	5	0	1	1	4
3 quart jug	0	3	0	3	1	1	0	3	0

Second solution

Vessel	Amount of wine in each vessel								
Stages	1	2	3	4	5	6	7	8	9
8 quart cask	8	3	3	6	6	1	1	4	
5 quart jug	0	5	2	2	0	5	4	4	
3 quart jug	0	0	3	0	2	2	3	0	

In the New York quarterly *Scripta Mathematica*[6] a British mathematician, Dr. W. W. Sawyer, gives a clever discussion of this kind of puzzle. He arrives at a general rule of procedure for their solution. Suppose we consider the case just discussed (8, 5, 3). Rule 1. If the jug 5 is empty, fill it from cask 8. Rule 2. If the jug 5 is not empty, there are two possibilities: a. If jug 3 is not full, fill it from jug 5; b. if jug 3 is full, empty it into the cask 8. Repeated application of this procedure leads to the desired result. The second solution above conforms to this rule. It may be tried on situations like (24, 17, 7), (12, 7, 5), and many others that can be imagined.

The Russian mathematician Y. I. Perelman prefers to have his work done for him by a robot which he affectionately calls his "clever little ball." The job of the robot is to run errands on a billard table having the unusual form of a parallelogram with a 60° angle. Here is how it works.

In order to solve the problem, considered above, with
the three vessels a = 5, b = 3, c = 8 we construct a parallelogram with sides OA = 5, OB = 3 and making an angle of 60°

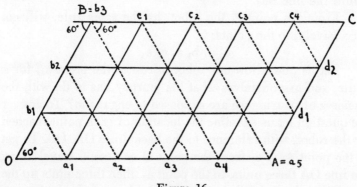

Figure 16

(Fig. 16). The robot runs his errands on this "billiard table,"
scrupulously observing the law of reflection, that is, the angle
of reflection is to be equal to the angle of incidence (Fig. 17).
Now suppose the robot is launched from the point O along the

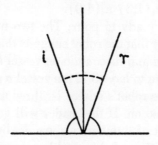

Figure 17

line OB. His first call point is $B = b_3$. From there, obeying the
aforementioned law, he will go to the point a_3. For the line
OB strikes BC at an angle of 60°, hence the robot will have to
continue on a line which also makes with BC an angle of 60°
This required line coincides with the diagonal of the

little rhombus adjacent to the lines OB, BC, since this diagonal bisects the angle OBC$=120°$ of that rhombus, by a known proposition of plane geometry. Thus the robot will follow the line Ba_3.

From the point a_3 the robot, by the same rule, will go, successively, to the points:

$$c_3 \; d_1 \; b_1 \; a_1 \; c_1 \; a_4$$

Let us now allow our robot to rest at the point a_4, for a while, and ask ourselves what his journey has to do with the business of pouring we are supposed to engage in? To answer the question let us examine the location of each point reached by the robot, with reference to the basic lines OA, OB. To get to the point, say, c_3 from the point O one could travel along the line OA three units, to the point a_3, then three units up the parallel to the line OB. Let us record this result in the form c_3 (3, 3). In professional mathematical parlance the numbers 3, 3 are said to be the coordinates of the point c_3 with reference to the basic lines, or axis OA, OB. Let us now rewrite the successive points of call of the robot, with the coordinates associated with each point. Thus: $b_3(0,3)$ $a_3(3,0)$ $c_3(3,3)$ $d_1(5,1)$ $b_1(0,1)$ $a_1(1,0)$ $c_1(1,3)$ $a_4(4,0)$.

Now we are ready to pour. The two numbers alongside the point b_3 signify that the robot suggests that you should have zero in the vessel a and three units in vessel b. For the point a_3 the robot wants you to have 3 in the vessel a and nothing in the vessel b. For c_3 the robot's advice is: three units in each of the vessels a, b. And so on. If the reader will go to the trouble of comparing the list of numbers we are dealing with right now with the first solution of this problem already given before, he may be surprised to find that the two sets of numbers are identical, or in other words, the dumb (?) creature has duplicated our solution. The reader who would amuse himself trying to work the same problem but launching the robot from O along the axis OA, has another surprise in store for him. The surprise is (or perhaps it is no surprise at all) that by his effort he will have found the second of the two solutions given before.

It is helpful to observe that the robot in its travels moves on two kinds of lines: 1°. lines parallel to the basic lines OA, OB; 2°. on those diagonals of the little rhombuses which bisect the angles of 120°. As a general rule those two kinds of lines are followed by the robot alternately. An exceptional case will be pointed out later.

The terminal point of the route solving a given problem may be marked beforehand. For instance, in the problem considered we want to have four gallons in each of two vessels, hence the problem will be solved when a contains that amount, that is, when the robot gets to the point $a_4(4,0)$.

It may also be observed that the robot is capable of providing two solutions of each problem: one when launched along the axis OA, and the other when launched along the axis OB.

The reader may try to work the problem: $a = 7$, $b = 5$, $c = 12$; $a = 9$, $b = 7$, $c = 16$. In the first problem the robot launched along the axis OA passes successively through the points:

$$a_7 \; c_2 \; a_2 \; b_2 \; d_2 \; c_4, \; a_4 \; b_4 \; d_4 \; c_6 \; a_6;$$

While following the peregrinations of our robot we tacitly assumed that the agile runner is provided with a mechanism which stops this servant when the preassigned point is reached. The question arises: what will happen if that braking mechanism should get out of order and fail to stop the robot, say, at the point a_4 in the second solution of the problem $a = 5$, $b = 3$, $c = 8$? Well, the faithful servant will simply continue to run in its unfailing obedience to the prescribed law and return back home, that is, to the starting point O, after having followed the supplementary path:

$$a_4 \; c_1 \; a_1 \; b_1 \; d_1 \; c_3 \; a_3 \; b_3 \; 0.$$

If we examine the round trip of the robot we notice that the path covered involves the points a_1 a_2 a_3 a_4 a_5. That is to say, the vessel a at one time or another contained the amounts of liquid 1, 2, 3, 4, 5. This is the very important discovery, indeed. For it shows that we could solve the problem, no matter what amount of liquid it would be required to have in the vessel a, not exceeding 5, of course. We thus enlarge the scope of our problems and enhance the value of the "robot method" of solving them. Thus in the suggested problem $a = 7$, $b = 5$, $c = 12$ we may ask to have in the vessel a three gallons of liquid. If we continue the list of "robot points," given above, beyond the point a_6 we find:

$$c_1 \ a_1 \ b_1 \ d_1 \ c_3 \ a_3 \ b_3 \ d_3 \ c_5 \ b_5 \ d_5 \ 0$$

Thus the desired point a_3 has been reached and our demand is satisfied. On the other hand if we wanted four units in the vessel a we would not have to go even as far as a_6. Let us notice, in passing, that here, too, we can have in a any amount between 1 and 7.

But is that always the case? Consider the problem: $a = 6$, $b = 4$, $c = 10$. In the usual way we obtain the following table:

	a_6	c_2	a_2	b_2	d_2	c_4	a_4	b_4	0
$a = 6$	2	2	0	6	4	4	0	0	
$b = 0$	4	0	2	2	4	0	4	0	
$c = 4$	4	8	8	2	2	6	6	10	

This table shows that in none of the three vessels can we have an odd number of gallons, and this includes the crucial number 5 which would divide the contents of c into two equal parts.

The examples which we have considered up to now have a common feature, namely that $c = a + b$. If c is greater than $a + b$, our "robot" method of solving the problems remains applicable. The "billiard table" remains unaltered, that is, its dimensions remain a, b, as before.

As an illustration let us solve the problem: $a = 6$, $b = 4$, c = 12. Our faithful robot may surprise us by following the same itinerary as in the preceding problem a = 6, b = 4, c = 10, with the same dire consequences, as the reader may readily verify.

Is our robot going to play on us the same trick whenever c is greater than a + b? To find the answer take the case when a = 6, b = 4, c = 11. It turns out that the first three lines of the solution of the problem a = 6, b = 4, c = 10 given above remain valid for our present case, with the same limitations for the vessel a (and b) as in that case. But the fourth line comes to the rescue. That line is now:

$$c = 5\ 5\ 9\ 9\ 3\ 3\ 7\ 7$$

Thus in the present case we can pour off not only any even amount but any odd amount as well. Any such odd amount

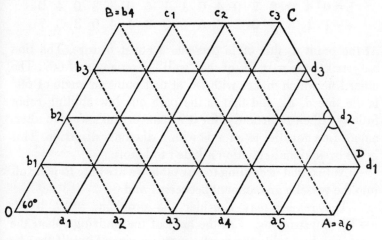

Figure 18

is none the worse for being in the vessel c than it would be in either of the two other vessels. Even the amount of 1 gallon can be had in c by filling the other two vessels.

The reader may investigate the cases when $a = 6$, $b = 4$,

and $c = 13, 14, 15, \ldots$ and try to formulate some rule about the results.

When c is *smaller* than $a+b$ the situation changes considerably. We may still have the services of the robot, but our billiard table has to be altered, namely, it has to be subjected to an amputation of a corner. Consider, for instance, the case when $a = 6$, $b = 4$, $c = 7$. In order to have our table we draw, as before, the axis $OA = 6$, $OB = 4$ at an angle of 60°. Now on the parallel to OA through B lay off $BC = 7 - 4 = 3$, and on the parallel to OB through A lay off $OD = 7 - 6 = 1$. The "wall" around our table is the broken line $OADCB$ (Fig. 18).

If we launch our robot from O along the axis OA, we obtain the following array:

	a_6	c_2	a_2	b_2	d_2	a_5	c_1	a_1	b_1	d_1	c_3	a_3	b_3	d_3	a_4	b_4	0
$a=6$	2	2	0	5	5	1	1	0	6	3	3	0	4	4	0	0
$b=0$	4	0	2	2	0	4	0	1	1	4	0	3	3	0	4	0
$c=1$	1	5	5	0	2	2	6	6	0	0	4	4	0	3	3	7

At the point d_2 this table presents a novel feature. The line $b_2 d_2$ strikes the part DC of the "wall" at an angle of 60°. The other line which makes with DC at that point an angle of 60° is the line $d_2 a_5$, and that is the path our law abiding robot follows, albeit reluctantly, for it is that servant's wont to alternate a line parallel to an axis with a diagonal direction. This is the exceptional situation alluded to before.

Notice that according to our table we are able to pour off into the vessel a any amount between 1 and 6.

The reader may consider the problem: $a = 9$, $b = 7$, $c = 12$. Constructing, with the help of the untiring robot, the corresponding table the reader may convince himself that he may have in the vessel a any amount between 1 and 9, with the conspicuous exception of 6, precisely the amount we would need if we wanted to divide the contents of c into equal parts. Sheer spite!

In the case $a = 6$, $b = 3$, $c = 9$, can we have in b one gallon or two gallons? If you ask the robot, the answer will be: No.

Many other intriguing problems of this sort may be solved by an interested reader, with the help of the robot. With a little practice this may become as good a game of solitaire as any.

D · The "False Coin" Problem   To the three "folklore" problems we have considered a good many more could be added, just as clever, just as ingenious. They all are a legacy of the past, a precious inheritance bequeathed to us by the centuries gone by. But what about the present century? Have the generations that have come before us been so much superior that they left to us a gift which we can enjoy but cannot duplicate, not to say rival? This is not the case. Our mathematical journals often publish questions fully as enticing as those we have learned from the folklore of the past.

Of the many examples that could be given let us take one that originated about the middle of the present century, in the United States.

In the issue for January 1945 the *American Mathematical Monthly* proposed the following question:[7] You have eight similar coins and a beam balance (without weights) At most, one coin is counterfeit and therefore lighter. How can you determine whether there is an underweight coin, and if so, which one, using the balance only twice?

The difficulty lies in the restrictive phrase, "using the balance only twice." Otherwise all one has to do is to put one coin on the balance and weigh against it each of the remaining coins, one at a time. If luck is against you, you may have to do it seven times before you arrive at the required answer. Being restricted to two weighings makes the difficulty. But the same is the case, say, with the wolf-goat-cabbage riddle.[8] There would have been no problem if the boat could accommodate all three, or even two of them. It is the restriction of "one passenger at a time" that creates the problem.

One who would give this coin problem a little thought is likely to agree that it is fully as challenging as any of the river crossing problems.

Eight months after the question appeared in the "Monthly" this periodical published a solution[9] which may be stated as follows: Weigh any three of the given coins against any other three of them. If the two sets balance, weigh the remaining two coins against each other, and the lighter of the two is the wrong coin. If the first two sets do not balance, the lighter of the two sets includes the suspect. Balance any two of those three coins against each other, and keep the third one in your hand. If they do not balance, the lighter coin is the one sought, and if they do balance, you are holding the culprit in your hand.

Soon after the appearance of this solution in the *American Mathematical Monthly*, the quarterly *Scripta Mathematica* published the following more general problem:[10] How, by balancing coins on a scale only three times can one detect which one of 12 apparently equal coins differs slightly in weight, without even knowing whether it is under- or over-weight?

Two solutions accompany the problem.

A few months later the *American Mathematical Monthly* proposed the same problem to its readers.[11] Two solutions were offered later.[12]

The solution given below follows closely the second solution in *Scripta Mathematica* mentioned above.

Throughout the discussion of the problem it is essential to bear in mind that only *one* of the coins is abnormal. Consequently, if two of the 12 coins are of equal weight, *both* coins are normal. More than that, if two groups of coins equal in number are also equal in weight, *all the coins in both groups are normal*.

To facilitate the solution of the problem it is necessary to make the coins distinguishable from one another by some kind of mark or marks. We shall assume that they have been numbered from 1 to 12.

Let us begin by putting four of the coins, say, 1234 in one pan of the scales, and four more coins, say 5678 in the other pan. Of this first weighing we consider the two alternatives: I: $1234 = 5678$; II: $1234 \neq 5678$.

In case I all the eight coins on the scale are normal, and the suspect is one of the coins 9, 10, 11, 12. Put any three of the latter coins, say, 9, 10, 11 in a pan, and in the other pan put any three coins we already know to be normal, say, 123. In this second weighing we consider the two possibilities:

Ia: $123 = 9, 10, 11$; Ib: $123 \neq 9, 10, 11$.

In case Ia the coin 12 is obviously the abnormal one, and a weighing, the third of the series, of the coin 12 against any one of the normal eleven coins will determine whether 12 is under- or overweight.

In case Ib one of the three coins 9, 10, 11, is abnormal, and is light or heavy according as the pan containing 9, 10, 11 is lighter or heavier than 123. In order to identify the wrong coin we take any two of the three coins under suspicion, say, 9, 10, and put them on the two different pans of the scale. If, in this third weighing of the series, it turns out that $9 = 10$, the abnormal coin is 11, and whether light or heavy we know already from the previous remark. If, however, $9 \neq 10$, one of these two is the culprit, and it is the lighter of the two, since we are looking for a lighter coin. This completes the discussion of case I.

Let us turn to case II. We observe, in the first place, that the abnormal coin is on the scale, and therefore the coins 9, 10, 11, 12 are normal. Furthermore, any one of the four coins 1234 may be either standard or heavy, but none of them is too light, while any one of the coins 5678 is either standard or light, but none of them is heavy.

For our second weighing let us put into one pan any

three of the four standard coins, say, 9, 10, 11 and add to them one of the possibly lighter coins, say, 5. On the other pan let us put two of the possibly heavy coins, say, 34, and let us add to them two of the remaining three possibly light coins, say 67. In this second weighing we consider the three possibilities:

$$\text{IIa: } 9,10,11,5 = 3467; \text{ IIb: } 9,10,11,5 > 3467:$$
$$\text{IIc: } 9,10,11,5 < 3467$$

In case IIa all the coins on the scale are standard, and so is the coin 12. Of the remaining coins 8 may be light, or of the coins 1, 2 one may be heavy. To decide the issue we put the latter two coins on the two pans of the scale. This third weighing may present us with one of the two possibilities $1 = 2$, $1 \neq 2$. In the first case 8 would be light. In the second case the heavier of the two coins is overweight.

IIb. That second weighing shows that 5 is necessarily standard and that one of the remaining two lightweight suspects 6, 7 is the guilty party. The issue can be decided by putting the two coins in the two pans of the scale (third weighing). The lighter of the two is the coin sought.

Case IIc could be brought about by 5 actually being light, or by one of the two coins 3, 4, both open to the suspicion of being heavy, actually being overweight. Put those two coins in the pans on the scale (third weighing). If they balance, then 5 is light; if they do not, the heavier is overweight.

The entire discussion is condensed in the chart which follows:

Case I

Weighing 1	Weighing 2	Weighing 3
	$\begin{cases} 123 = 9,10,11 \\ \\ 123 \neq 9,10,11 \end{cases}$	$\begin{matrix} 12 \\ 9 = 10 \dots\dots 11 \\ 9 \neq 10 \end{matrix}$
$1234 = 5678$		

Case II

$$1243 \neq 5678 \begin{cases} \text{a. } 9,10,11,5 = 3467 \begin{cases} 1=2 & \text{8 is light} \\ 1 \neq 2 & \text{The heavier is over-} \\ & \text{weight} \end{cases} \\[2em] \text{b. } 9,10,11,5 \neq 3467 \begin{cases} 6=7 & \text{8 is light} \\ 6 \neq 7 & \text{The lighter of the} \\ & \text{two is light} \end{cases} \\[2em] \text{c. } 9,10,11,5 \neq 3467 \begin{cases} 3=4 & \text{5 is light} \\ 3 \neq 4 & \text{The heavier of the} \\ & \text{two is overweight} \end{cases} \end{cases}$$

The interest in the "bad coin" problem did not remain confined to this continent. Right from the start the problem attracted the attention of the *Mathematical Gazette* of London. As early as 1945 this periodical published a note on it.[13] Two other notes were published by the *Gazette* the following year, 1946,[14] and in 1947 the *Gazette* devoted to the question a nine page article.[15] Among the questions discussed is the following: We have seen that the twelve coin problem has been solved in several different ways. The question may therefore be asked: Given a number of coins and the number of weighings necessary for the detection of the counterfeit coin, in how many different ways may this problem be solved?

There are many other ramifications that have been suggested or considered in connection with this problem; what has been said may perhaps have given an idea in what way and with what speed a riddle of this sort expands and develops nowadays.

Has this problem been exhausted? One of the three contributors to the volume of the *Scripta Mathematica* for the *year 1946*, to which reference was made, seemed to be quite sure that this is the case, judging by the fact that he entitled his article: *Epitaph on the coin problem*. It would seem, however, that he spoke out of turn. For the same *Scripta Mathe-*

matica in its issue for March—June 1950 published an article under the heading: "Those twelve coins again."

The well-springs of wisdom of the human race have not gone dry. They are more active, more vigorous than they have ever been. This manifests itself just as surely in mathematics on the recreational level as on the most abstract heights that mathematics can reach.

2 · FAMOUS PROBLEMS

Introduction   Some problems are famous and deserve special attention. Felix Klein's (1849-1925): *Lectures on Famous Problems of Elementary Geometry*," namely the trisection of an angle, the squaring of the circle, and the duplication of the cube[16] has been a classic for many years. Other such collections have been published more recently.[17]

Now suppose we agree that there are famous problems in mathematics. What then is a famous problem? It is by no means easy to answer this question. But we can make a few guesses and see where they may land us.

Is a "famous" problem one of long standing, an ancient problem? Well, yes. There are such problems, and the problems of antiquity which were just mentioned above are the best examples of the kind. There are problems that are even older than those, as for example, the "river crossing" problem discussed in the preceding section.[18]

A · Morley's Problem   But venerable age is by no means an indispensable characteristic, or, as we say in mathematics, a "necessary" condition for a "famous" problem. There are problems as famous as any and which originated in the present century. One of them is the late Frank Morley's trisector problem. Morley (1860-1937) was not concerned with the methods of trisecting an angle. He took that part of it for granted. He trisected all three angles of a triangle and took the points of intersection of the pairs of trisectors adjacent to the three sides of the triangles. The triangle formed by the three points thus obtained is, according to Morley, equilateral.

Morley arrived at this property, while studying some higher plane curves, during the first decade of the present century. The problem became famous almost overnight. It ended up by attracting the attention of men of the stature of Henri Lebesgue (1875-1941).

The proposition was proved by many writers in a variety of ways. Moreover, Morley's contemporaries exhibited a great deal of ingenuity in generalizing his simple theorem. Those mathematicians first considered all the points of intersection of the six trisectors, picking out more equilateral triangles, then they replaced the trisectors by n-sectors of the angles, etc. The original prototype is completely snowed under, and the problem as a whole has by now an enormous bibliography. You will grant that the problem is entitled to be considered famous.

B · The Problem of Apollonius Does a "famous" problem have to be attached to a famous name? Certainly, this is a valuable asset for a problem. If famous names in the world of the movies or the world of sports may be used to tell you which cigarettes you should smoke, what beer to drink, what soap to wash your hands with, why would not a famous name in the mathematical world be a good recommendation for a problem in mathematics? There are many such problems. Thus the problem of drawing a circle tangent to three given circles is known as the problem of Apollonius, who is supposed to have solved it. But his solution did not come down to us. François Viète (1540-1603) was the first to produce a solution. René Descartes (1596-1650) and Isaac Newton (1642-1727) worked on this problem. Is this not enough to entitle any problem to fame?

The problem rose again to the heights of prominence when in the early nineteenth century the newly developed or discovered theories of the radical axis, centers of similitude, etc., were applied to its solution. It figured prominently in the competition between the analytic and synthetic methods in geometry of that time. The problem was used as a proof that constructions to which analytical considerations give rise may be as simple as those obtained by purely geometrical reasoning. These arguments, not to say quarrels, involved men as illustrious as Poncelet, Gergonne, and others.

C · "Fermat's Last Theorem"   None of the famous problems outranks "Fermat's last theorem," namely the proposition that the equation

$$x^n + y^n = z^n$$

has integral solutions in x, y, z for $n = 1$, 2, and for no other number greater than 2. An incredible amount of time and effort on the part of mathematicians, from the lowliest to the most renowned, has been devoted to this problem. Some mathematicians, like H. S. Vandiver (1882—), made this problem the center of their activities in research. During the second half of the nineteenth century a prize was established to be awarded to any one who could prove or disprove Fermat's last theorem. The prize was no trifle, for it amounted to one hundred thousand German marks, the equivalent at the time of $25,000. At the time when the prize was established most college professors, both in Europe and in America, could not expect to earn much more in a dozen years of teaching. It is reasonable to assume that many an ambitious soul had his eyes fixed on this carrot. But you need not worry about it. Whether you try or not, you cannot get that prize any longer. Not that somebody got ahead of you. The inflation and deflation which followed the first world war rendered the bequest worthless.

The problem remains unsolved. It is the more tantalizing

due to the claim of Fermat that he had a very short and simple proof for it, but he kept it to himself, so that others could have the pleasure of discovering it in their turn.

D · Goldbach's Conjecture Christian Goldbach (1690-1764) was a contemporary of Euler. He was Russian and lived in St. Petersburg (now Leningrad). One day he made a shrewd guess—in dignified scientific language it is called a conjecture—that every even number is the sum of two prime numbers. Thus $10 = 3 + 7$, $24 = 11 + 13 = 7 + 17$.

Goldbach communicated his guess to his illustrious friend Leonhard Euler (1707-1783), who was quite impressed. The surmise seemed to him to be a true proposition. But his sustained efforts to prove it were all in vain. And so were the efforts of all those followers of his who tried their hand at it, up to the present time, or almost.

In the late thirties of the present century the Russian mathematician I. M. Vinogradov proved that any odd number is the sum of three prime numbers. Thus $21 = 3 + 7 + 11$, $35 = 5 + 7 + 23 = 7 + 11 + 17$. In connection with Vinigradov's proof E. T. Bell (1883—) remarks that the work of the Russian inspires sympathy for Euler, in his failure to prove Goldbach's surmise. In the wake of Vinogradov two other Russian mathematicians, Linnik and Tchudakov, produced other proofs of Vinogradov's theorem.

Does Vinogradov's theorem prove Goldbach's surmise? Sufficiently so to warrant the change of the name from Goldbach's surmise to the Goldbach-Vinogradov theorem, but not enough to consider the question as quite settled.

A great many other famous problems associated with great names may be added. Nevertheless this is not a necessary requirement for a problem to be famous. The problems of antiquity may again serve as an appropriate example. The "coin" problem[19] if its fame should turn out to be of the lasting variety, is likely to continue to be known just as the "coin" problem, and nothing more.

E · The Problem of the Tangent   The famous problems that we have considered so far are tainted with a degree of glamor for one reason or another. But it cannot be said that glamor is an indispensable attribute of a famous problem. There are famous problems which can hardly lay claim to being spectacular, glamorous, as, for instance, the problem of drawing a tangent to a curve at a point on the curve. On the other hand, the whole history of mathematics is, in a way, reflected in the history of this problem.

The Greek definition of a tangent is: a line passing through the given point and such that no other line drawn through the point can lie between the tangent and the curve. It is a far cry from this definition to the definition according to which a tangent is the limiting position of a secant, the definition which is now standard in our textbooks.

The various ways in which the solution of the problem of drawing a tangent to a curve is solved are intimately connected with the way we define the term "curve." "Tell me what a curve is, and I will tell you what a tangent is." If we consider, with Roberval (1602-1675), whose actual name, by the way, was Gilles Personnier, that a curve is the path of a moving point, the tangent is determined by the parallelogram of velocities. If you define a curve as a graph of an analytic equation, as we do now when we use analytic methods, the direction of the tangent to the curve is determined by the derivative of the function considered. In Projective geometry the definition of the tangent is an immediate consequence of the mode of generation of the curve considered.[20]

The tangent, or rather its absence, came again prominently to the fore when in 1861 Karl Weierstrass (1815-1897) made the extraordinary and unbelievable discovery that a continuous (that is, a smooth, unbroken) curve may not admit of a tangent at any of its points. The story of the problem of the tangent is neatly summarized by Paul Serret in his little book "Des Méthodes en Géométrie."[21]

F · *The Recurrence of "Famous" Problems* We have seen that a famous problem may be very old, or more recent, and even very recent; it may or may not be attached to the name of a great mathematician, that it may or may not be of intrinsic importance in the history of mathematics. There are, however, some characteristics which seem to be fairly common to all of them. All of them are simple in their statement, readily make a picture in your mind, and can therefore be carried in your memory without effort. In brief, a famous problem has a "simple formulation."

But the best definition that could be given of a famous problem is the following: "A famous problem is one that nobody ever heard of, least of all those who busy themselves with it."

This sounds paradoxical, but it is the inevitable conclusion one arrives at when one examines the history of such problems, and the more famous the problem, the more applicable is the paradox. But it is not so strange, after all, if you think about it. It is inherent in the very nature of those problems.

Since those problems are simple in their conceptions, they occur to many individuals independently, without knowledge of any previous efforts by others in connection with any one of those problems. In past centuries this was aggravated by the paucity of means of communication between individual mathematicians. The same effect is produced in modern times by the multiplicity of mathematical journals published in many different countries, in a number of different languages too large for comfort.

G · *Conclusion* One may raise the question: Of what use is all this time and effort spent so lavishly on these questions? Who will ever have any need for the answers provided? Well, these are tough questions. One may have no qualms in telling you that nobody is expected to make any "use" of the results obtained in most of these problems.

Mathematics is useful. It is the practical need for mathematics that accounts for its origin. The continuous development of mathematics is due to the constantly growing need for answers to questions that mathematics can supply.

It is nevertheless true that the part of mathematics that has practical, bread and butter applications is a small part of the whole body of the science. Some of this residue may yet become useful one day. But the bulk of it may never be. Do you think that the so called "perfect numbers" could have great practical value? Who will argue that the two new perfect numbers which were recently computed are destined to play a great role in human affairs, in spite of the hundreds of digits it would take to display each of them in its full magnificence? But if they are not of any use, why waste the wonderful computing machines on such futilities? The answer is, because it is interesting, because the riddle of the perfect numbers is a challenge which we would like to meet.

Most of mathematics is in the same boat. You may say that the inquiring genius of man has erected the stately edifice which is mathematics as a monument to the greatness of his intellectual power, as a permanent proof that "man cannot live on bread alone." If you are not given to selfglorification and "highsticularious" language, you may simply say that a man is curious and is willing to pay a high price, in time and in effort, to find an answer to what strikes his fancy. He likes to exercise his inventiveness and to display it before others, namely before those who can get as excited about it as he does himself.

That is the best that can be said in defense of most of mathematics. If one cannot derive a world of satisfaction from the solution of a problem, if one cannot be charmed by the result obtained, regardless of the ulterior values that result may or may not have, one is not a mathematician. This is the outstanding fact exhibited by the famous problems. They offer a challenge, they present an opportunity for a display and an exercise of cleverness, of intellectual prowess. May those enjoy it who can.

3 · WITHOUT THE BENEFIT OF PAPER AND PENCIL

A · Mathematics and Computation   In the minds of a great many people the terms "mathematics" and "calculations" are synonymous. To be a "great mathematician" is to be a rapid computer. This idea is superficial. It is true that some of the great mathematicians were also skillful and accurate computers. But those were the exceptions. As a rule rather the contrary is the case. To take but one example, let us quote a statement made by Henri Poincaré, one of the greatest mathematicians that ever lived: *"Quant à moi, je suis obligé de l'avouer, je suis absolument incapable de faire une addition sans faute."* (As to myself, I am obliged to own I am absolutely unable to perform an addition without making a mistake.)[22]

In some branches of mathematics calculations play a very minor, even negligible role. Any high school student, past or present, who was exposed to a course in plane geometry found that out for himself. But even when calculations are used in a mathematical problem, they are seldom the core of the matter. An artist in the process of producing a painting has to mix his paints. Some painters acquired a considerable renown for having devised very effective shades of some colors, like the "Titian red." But it does not occur to anybody to identify the profession of an artist with paint-mixing.

If a mathematician has to answer a question which calls for a number, he may have to do some computation to obtain the required result. However, the essential part of the solution of the problem is not the computations, but the reasoning process which enables the solver to *choose* the appropriate computations. It is this intellectual effort of analyzing the situation that constitutes the mathematical character of the problem. The wolf-goat-cabbage riddle and its generalizations[23] are a good example very much to the point. As an illustration less ancient but even more forceful one may quote the two elegant solutions devised by Pascal and Fermat of the problem proposed by the Chevalier de Méré.[24]

The "eight coin" problem[25] requires analysis and imagination, or wit and shrewdness, if you prefer simpler words, but no writing is needed. It would be difficult to handle the "twelve coin" problem[26] completely without writing, on account of the considerable number of different cases involved that it is necessary to keep track of. But the writing is not what makes for the added interest of the problem as compared with its simpler prototype. Of course, one may object that those are riddles rather than mathematical problems in the usual sense of the word. But this is rather in favor of the contention presented here than against it. The fact that such things appear in professional mathematical journals and are grappled with by the readers of those publications goes to show where the mathematician feels his meat is.

The problem of the suburban traveler which we considered before[27] is more of the standard type and requires some calculations, but those are of the kind that can be performed mentally. The achievement of the solver does not lie in the arithmetical skill shown, but in the intellectual acumen used to grasp the interrelations involved. The following problems are other examples of this kind. The reader may find satisfaction in attempting to solve them and check his solutions against those given below. Some of those problems may serve to intrigue and enliven many a social gathering.

B · Problems   a. If each boy at a picnic were given three apples from the available supply, one of the boys would have to be satisfied with two apples. But if each boy were given two apples, eight apples would remain. How large was the supply of apples?

b. Two trains start out at 7 A. M., one from A going to B and the other from B going to A. The first train makes the trip in 8 hours and the second in 12 hours. At what hour of the day will the two trains pass each other?

c. Three brothers, Tom, Dick, and Harry, stopped at an inn and ordered a dinner. When at the end of the meal no dessert was served, they asked the innkeeper to stew some prunes for them. While waiting in their comfortable chairs for the prunes, all three of them fell asleep. After a while Tom woke up and, finding a bowl of prunes on the table, ate his share and went back to sleep. When Dick woke up a little later, he ate what he thought was his share and he, too, fell asleep again. When Harry awoke he proceeded the same way. When Tom awoke for the second time, he aroused the two younger brothers, and after a little discussion the whole story was cleared up. The remaining eight prunes were divided equitably between Dick and Harry. How many prunes did each of them get?

d. A steamer plying between two river ports A and B makes the trip from A to B in 12 hours and the return trip in 18 hours. How long will it take a log thrown into the water at A to reach B?

e. Find two numbers whose difference and whose quotient are both equal to three.

f. A commuter ordinarily reaches the railroad station nearest his home at 5 P. M., where he is met by his wife, in the family car. One day he unexpectedly arrived at the station at 4 P. M. and instead of waiting for his car at the station he started out for home, on foot. After a certain length of time he meets his wife and makes the rest of the way home in the car, as usual. He reached home sixteen minutes ahead of the usual time. How long did he walk?

g. Find a number such that if one sixth part of it is

multiplied by one eighth part of it, the result is equal to the number.

h. A summer camper rowed one mile up-stream when his hat blew off into the water beside him. As it was an old hat he decided to let it go. Ten minutes later he remembered that he put his return ticket under the hatband. Rowing at the same rate as before, he reached the hat (and the ticket) at the same point where he started out in the boat. How fast was the stream flowing?

k. What is the smallest number of cuts that would divide a cube of wood 3 inches on the edge into cubes one inch on the edge?[28]

m. A courier rode from the rear of a column of marching soldiers to the front and returned forthwith to the rear of the column. He kept his horse jogging along exactly three times as fast as the column itself was advancing. Where on the road, with reference to the original position of the vanguard, did he complete his journey?

C · Solutions   a. Each boy will have two apples if we take back one apple from each lot of three apples planned originally, and we will thus accumulate eight apples. Hence the supply of apples consists of $8 \times 3 + 2 = 26$ apples.

b. First solution. The speeds of the two trains are inversely proportional to the time it takes them to cover the distance AB, hence the ratio of those speeds is equal to $12 : 8 = 3:2$, and the distance the two trains cover in the same length of time are proportional to their speeds, that is $3 : 2$. At the time the two trains meet the train from A and the train from B will have covered, respectively, 3/5 and 2/5 of the distance AB. The time the train from A traveled to reach the meeting point is 3/5 of 8 hours, that is, 4 hours and 48 minutes, so that the trains meet at 11.48 A. M.

Second solution. The trains cover, respectively, 1/8 and 1/12 of the distance AB, per hour. Hence they approach each other by a $1/8+1/12=5/24$ part of the distance AB per hour. Thus they will meet after 24/5 hours of travel.

c. First solution. Tom left 2/3 of the number of prunes he found on the table. Dick left 2/3 of the prunes he found, which was $2/3 \times 2/3 = 4/9$ of the number of prunes the innkeeper served. Finally Harry left $4/9 \times 2/3 = 8/27$ of the original number of prunes, and this amounted to eight prunes. Hence the innkeeper served originally 27 prunes in all. Of those Tom ate 9, Dick ate 6, and Harry 4. Thus of the eight prunes remaining Dick is entitled to $9-6=3$ and Harry to $9-4=5$.

Second solution. For this "frontal attack" upon the problem we may substitute a "back door" solution. Harry left 8 prunes for his two brothers, hence he ate himself four prunes, that is, he found on the table $8+4=12$ prunes. That was the number Harry left for the other two brothers, hence he found on the table 18 prunes, left by Tom, etc.

Notice that the arithmetic is about fourth grade level, but the problem is not.

d. The hourly rate of the steamer going up-stream is less than the hourly speed going down-stream by two hourly speeds of the current. After having traveled 12 hours *up-stream* the boat is therefore $2 \times 12 = 24$ hourly speeds of the current away from its destination A. This distance the boat is expected to cover in $18-12=6$ hours, hence, when going up-stream, the boat covers per hour a distance equal to $24 : 6 = 4$ times the hourly speed of the current, and in 18 hours the boat covers $18 \times 4 = 72$ such distances, which is thus the number of hours it will take the log to cover the distance AB.

e. The quotient of the two numbers being 3, the difference between the larger number and the smaller number is equal to twice the small number. On the other hand, this difference is equal to three, hence the smaller number is equal to 3/2, and the larger number to $3/2+3=9/2$.

f. Instead of worrying about the man, it is more to the

point to consider the role of the wife (As usual: *Cherchez la femme!*). She, too, saved 16 minutes on her usual trip. This time is made up of eight minutes saved by not going from the meeting place to the station and of eight minutes saved on the return trip. But the wife expected to be at the station at 5 P. M., hence the meeting took place at 4.52 P. M., and "hubby" thus exercised for 52 minutes.

Notice that the lady in the car travels $52/8 = 13/2$ times faster than her husband travels on foot.

g. If instead of multiplying 1/6 of the required number by 1/8 of it we would multiply that number by itself, our result would be $8 \times 6 = 48$ times larger than expected, that is, it would be equal 48 times the required number. Thus multiplying the number by itself produces the same effect as multiplying it by 48, hence 48 is equal to the required number.

h. The important circumstance to notice in the situation is that when the camper rowed away from the hat and against the current, his rate of separation from the hat was equal to his rate of rowing in still water, and this is also the rate of approach to the hat when he turns around and tries to catch up with his hat. It follows that since he was rowing away from the hat for ten minutes, it will take him ten minutes to catch up with that precious hat (and the ticket in it). Thus the hat was in the water a total of twenty minutes and in that length of time covered a distance of one mile, hence the hourly rate of the current is three miles.

k. Assume that the given cube lies on a horizontal floor. We divide the top face of the cube into nine equal squares by two pairs of parallel lines. By four vertical cuts along those four lines we divide the cube into nine equal columns which can be divided into twenty seven equal cubes by two horizontal cuts. The assigned task has thus been accomplished by six cuts. Could it be done with a smaller number of cuts? The answer is: No. This becomes clear when we consider the small cube which occupied the center of the given cube: each of its six faces had to be obtained by a separate cut.

m. Since the courier travels at a speed equal to three times the speed of the marching column, he approaches the vanguard of the column on the first part of his journey with a speed equal to twice the speed of the column (the vanguard is moving away from him). Hence during the time the courier traveled from the rear to the head of the column, the column moved on a distance equal to half the length of the column.

On his return trip the courier approaches the rear of the column at a speed equal to $(3+1=4)$ four times as great as the speed of the column (the rear is coming to meet him). Hence by the time the courier reached the rear, the column traveled one fourth of its own length. Thus the courier returns to the rear of the column at a point on the road which is at a distance equal to $\frac{1}{2}+\frac{1}{4}=\frac{3}{4}$ the length of the column, and this point is also $\frac{1}{4}$ of the length of the column behind the point occupied by the vanguard when the courier started out on his journey.

FOOTNOTES

[1] Cf. Chapter V, Section 1e.

[2] Cf. Chapter VII, Section 3.

[3] Cf. Chapter II, Section 1a.

[4] "Digital Reckoning Among the Ancients," Leon J. Richardson, *American Mathematical Monthly*, Vol. 23, No. 1, Jan. 1916, pp. 7-13.

[5] Cf. Chapter I, Section 3c.

[6] *Scripta Mathematica*, March-June 1950.

[7] *American Mathematical Monthly*, January 1945, p. 42.

[8] Cf. Chapter VII, Section 1a.

[9] *American Mathematical Monthly*, August-September 1945, p. 397.

[10] *Scripta Mathematica*, Vol. 11, Nos. 3-4, Sept.-Dec., 1945, p. 360.

[11] *American Mathematical Monthly*, Vol. 53, No. 3, March 1946, p. 156.

[12] *American Mathematical Monthly*, Vol. 54, No. 1, January 1947, pp. 46-48.

[13] *Mathematical Gazette* (London, 1945), p. 227.

[14] *Ibid.*, 1946, p. 231ff.

[15] *Ibid.*, 1947, pp. 31-39.

[16] *Lectures on Famous Problems of Elementary Geometry*, Felix Klein, translated by W. W. Beman and D. E. Smith (sec. ed.), revised by R. C. Archibald (New York, 1930).

[17] Doerrie, Henrich, *Triumph der Mathematik*, the subtitle of which reads as follows: *Hundert beruemte Probleme aus zwei Jahrtausenden mathematischer Kultur* (one hundred famous problems from the mathematical culture of the last two thousand years), (Breslau, 1933). Callandreau, Edouard *Célèbres problèmes des Mathèmatiques* (Paris, 1949).

[18] Chapter VII, Section 1a.

[19] *Ibid.*, Section 1d.

[20] Cf. Chapter III, Section 3h.

[21] *Des méthodes en géométrie*, Paul Serret (Paris, 1855).

[22] *Science et Méthode* (Paris, 1909), p. 46.

[23] See Chapter VII, Section 1a.

[24] Cf. Chapter V, Section 2b.

[25] Chapter VI, Sec. 1d.

[26] Cf. Chapter VII, Section 1d.

[27] Chapter I, Section 3c.

[28] Cf. Chapter I, Section 3c.